Coevolution and Systematics

Proceedings of an International Symposium held in Brighton, July, 1985 sponsored by the Systematics Association as part of the Third International Congress of Systematics and Evolutionary Biology

The Systematics Association
Special Volume No. 32

Coevolution and Systematics

Edited by

A. R. Stone
*Nematology Department,
Rothamsted Experimental Station,
Harpenden, Herts., U.K.*

and

D. L. Hawksworth
*CAB International Mycological Institute,
Ferry Lane, Kew, Surrey, U.K.*

Published for the SYSTEMATICS ASSOCIATION by

CLARENDON PRESS · OXFORD

1986

Oxford University Press, Walton Street, Oxford OX2 6DP
Oxford New York Toronto
Delhi Bombay Calcutta Madras Karachi
Petaling Jaya Singapore Hong Kong Tokyo
Nairobi Dar es Salaam Cape Town
Melbourne Auckland
and associated companies in
Beirut Berlin Ibadan Nicosia

Oxford is a trade mark of Oxford University Press

Published in the United States
by Oxford University Press, New York

© The Systematics Association 1986

All rights reserved. No part of this publication may be reproduced, stored in a retrieval system, or transmitted, in any form or by any means, electronic, mechanical, photocopying, recording, or otherwise, without the prior permission of Oxford University Press

This book is sold subject to the condition that it shall not, by way of trade or otherwise, be lent, re-sold, hired out or otherwise circulated without the publisher's prior consent in any form of binding or cover other than that in which it is published and without a similar condition including this condition being imposed on the subsequent purchaser

British Library Cataloguing in Publication Data
Coevolution and systematics. —(The Systematics Association special volume; no. 32)
1. Phylogeny   2. Biology—Classification
I. Stone, A. R.   II. Hawksworth, D. L.
III. Series
574'.012   QH367.5
ISBN 0-19-857703-6

Library of Congress Cataloging in Publication Data
Coevolution and systematics.
(The Systematics Association special volume; no. 32)
"Proceedings of an international symposium held in Brighton, July 1985, sponsored by the Systematics Association as part of the Third International Congress of Systematic and Evolutionary Biology"—P.
Bibliography: p.
1. Coevolution—Congresses.   2. Biology—Classification—Congresses.   3. Host-parasite relationship—Congresses   I. Stone, A. R.   II. Hawksworth, D. L.
III. Systematics Association.   IV. International Congress of Systematic and Evolutionary Biology (3rd : 1985 : Brighton, East Sussex)   V. Series.
QH372.C64   1986   574'.012   86-5307
ISBN 0-19-857703-6

Typeset by Dobbie Typesetting Service, Plymouth, Devon
Printed in Great Britain by
St Edmundsbury Press Ltd, Bury St Edmunds, Suffolk

Preface

Interest in the coevolution of organisms has increased considerably in recent years as evidenced by a range of symposia, books, and review articles. The literature is, inevitably, highly dispersed and has a variety of emphases depending on the standpoint of the authors, or of organizations sponsoring particular meetings. This volume does not attempt to give a comprehensive review of coevolution on the lines of texts already available (e.g. D. J. Futuyma and M. Slatkin (eds) *Coevolution*, Sinauer Associates, Sunderland, Mass., 1983; O. Kraus (ed.) *Co-evolution* Sonderbande des Naturwissenschaften Vereins in Hamburg, Paul Parey, Hamburg and Berlin, 1978; I. Hedberg (ed.) *Parasites as plant taxonomists* Symbolae botanicae upsalienses 22 (4), 1979), but rather to present a series of essays which are related to the implications of coevolution phenomena for sytematics.

The subject has not previously been addressed by the Systematics Association, but the occasion of the Third International Congress of Systematics and Evolutionary Biology in Brighton, England, 5–10 July 1985 presented an opportunity and the Association organized a one-day symposium on Coevolution and Systematics as part of the programme. In convening this meeting it was our intention to invite a small number of specialists with different backgrounds to consider the implications of putative coevolutionary phenomena for the systematics of the organisms involved.

The most conclusive evidence for coevolution is held to be found in host–parasite relationships and other similarly close associations and, inevitably, the papers in this volume deal with such relationships. Barrett (Ch. 1) and Parlevliet (Ch. 2) draw on fungal pathogens of vascular plants; Eastop (Ch. 3) deals with phytophagous aphids; Humphries *et al.* (Ch. 4) with a tripartite relationship of fungi, moths, and a genus of vascular plants; Lyal (Ch. 5) and Beveridge (Ch. 6) with parasites of vertebrates; and Thompson (Ch. 7) provides an overview.

Thinking on the systematics of parasite groups, and to a lesser extent their hosts, has been much influenced by the assumption that coevolution between hosts and parasites has resulted in parallel traits in their phylogenies which, in turn, may be reflected in their systematics. A plethora of 'rules' has been proposed, of which the most important and succinct is Fahrenholz's rule, which may be paraphrased as 'parasites and hosts are associated by descent with consequently congruent phylogenies'.

In convening the meeting the editors attempted to avoid the influence of their own preconceptions about the utility of such rules, but in the proceedings published here we see a consistent expression of reservation about their applicability, which our personal experience with nematodes and fungi also supports. Much of the justification for the use of such rules has been on intuitive arguments, as is some of the evidence, for and against, presented here. However, we see in two papers (Lyal, Humphries *et al.*) assessments based on tests of parallelism in cladograms of hosts and parasites, an approach the rigor of which we do not believe can be denied. The conclusion from the relationships addressed in these contributions is that, while the parasitological rules have sound application in a small proportion of cases, as generalities they are much less justifiable and can be positively misleading.

The principal conclusion for practising systematists is to encourage them to pay careful attention to correlations between hosts and organisms dependent on them, critically examining cases where anomalies appear to occur. At the same time it is also clear that systematists must be cautious, must not overemphasize relationships as 'jumps' from one host to a rather distantly related one can sometimes occur, and should remember that their data matrix may be incomplete due to the inadequate collecting of often inconspicuous organisms.

Harpenden and Kew A. R. S.
December 1985 D. L. H.

ALAN RAMSEY STONE

After the final proofs of this volume had been returned to the press Dr Stone was found dead in his laboratory at Rothamsted on 6 May 1986. His unexpected death, at the early age of 43, came as a great shock to all who had come to value him as both friend and colleague. In addition to his distinguished contributions to nematode systematics and pathology Alan had become a key figure in the Systematics Association. He served as Zoological Secretary from 1977 to 1983 when he became Editor-in-Chief of the Special Volume Series, supervising its transfer to the Oxford University Press. His broad systematic interests are also reflected in the contribution he made to the Linnean Society of London, serving as a Vice-President in 1983–4, and to the CAB International Institute of Parasitology, of which he was Consultant Director from 1983. His premature death is a tragedy for his family, friends, and colleagues and for nematology and systematics in general.

Kew
11 June 1986   D.L.H.

Contents

	List of contributors	xi
1.	Host–parasite interactions and systematics	1
	John A. Barrett	
	Introduction	1
	Variation in virulence and resistance	3
	Host and parasite evolution	6
	Parasite specificity and host phylogeny	9
	Parasite specialization in natural ecosystems	12
	Discussion	13
2.	Coevolution of host resistance and pathogen virulence; possible implications for taxonomy	19
	J. E. Parlevliet	
	Introduction	20
	Coevolution of host and pathogen: a model	20
	Specialization	21
	Biological specialization and coevolution	23
	Coevolution	26
	Implications for taxonomy	31
3.	Aphid–plant associations	35
	V. F. Eastop	
	Introduction	35
	Host ranges of the major groups of aphids	37
	Host ranges in different geographical areas	45
	Biologies of aphids on different groups of plants	47
	Anomalous host–plant relationships	48
	Phylogeny of Aphidoidea	51
	Conclusions	52
4.	*Nothofagus* and its parasites: a cladistic approach to coevolution	55
	C. J. Humphries, J. M. Cox, and E. S. Nielsen	
	Introduction: parasitism by 'association by descent' and 'colonization'	55

	Nothofagus: the host phylogeny	61
	The parasites	62
	Historical biogeography	71
	Conclusions	73
5.	Coevolutionary relationships of lice and their hosts: a test of Fahrenholz's rule	77
	C. H. C. Lyal	
	Introduction	78
	Effectiveness of Fahrenholz's Rule	79
	Alternative hypotheses	83
	Systematic implications	87
6.	Coevolutionary relationships of the helminth parasites of Australian marsupials	93
	Ian Beveridge	
	Introduction	94
	Origins of the marsupials	95
	Origins of the parasite fauna of Australian marsupials	96
	Coevolutionary patterns	103
	Convergence	108
	Coevolutionary mechanisms at the species level	110
	Conclusion	112
7.	Patterns in coevolution	119
	John N. Thompson	
	Introduction	120
	Coevolution and systematics: Umbelliferae and insects	121
	Coevolution and speciation	127
	Limits to the arms race and evolution of mutualism	130
	Population structure and coevolution	134
	Interaction norms: a new term	135
	The merging of approaches	136
	List of Systematics Association Publications	145

Contributors

JOHN A. BARRETT
Department of Genetics, University of Cambridge, Downing Street, Cambridge CB2 3EH, U.K.

IAN BEVERIDGE
South Australian Department of Agriculture, GPO Box 1671, Adelaide 5000, Australia.

J. M. COX
Department of Entomology, British Museum (Natural History), Cromwell Road, London SW7 5BD, U.K.

V. F. EASTOP
Department of Entomology, British Museum (Natural History), Cromwell Road, London SW7 5BD, U.K.

C. J. HUMPHRIES
Department of Botany, British Museum (Natural History), Cromwell Road, London SW7 5BD, U.K.

C. H. C. LYAL
Department of Entomology, British Museum (Natural History), Cromwell Road, London SW7 5BD, U.K.

J. E. PARLEVLIET
Plant Breeding Department, Agricultural University, Lawickse Allee 166, 6709 DB Wageningen, The Netherlands.

E. S. NIELSEN
Division of Entomology, CSIRO, GPO Box 1700, Canberra ACT 2601, Australia.

JOHN N. THOMPSON
Departments of Botany and Zoology, Washington State University, Pullman, Washington, 99164, U.S.A.

1. Host–parasite interactions and systematics

JOHN A. BARRETT

Department of Genetics, University of Cambridge, U.K.

Summary

Using examples mostly from plant host–parasite systems, the assumption that there may be an *a priori* expectation that host and parasite phylogenies should exhibit congruence is critically discussed.

Introduction

Ecosystems are interlocking webs of interactions between species and, consequently, member species of natural communities should have an effect on and be affected by each other's evolution. The more intimate the association between species, the stronger the coevolutionary forces are likely to be. In associations as close as those between a parasite and its host, it would seem reasonable to expect that the evolutionary history of both host and parasite should be intertwined and that the taxonomy of the parasite should be reflected in that of the host and *vice versa*. In other words, host and parasite species should be 'associated by descent' (Mitter and Brooks 1983). More precise interpretations of the phylogenetic relationships between host taxa and between parasite taxa should be possible if the associations between the host and parasite groups are included in the analysis. This is what the parasitological method sets out to do. It can be viewed as a two-step process (cf. Humphries *et al.*, and Lyal, this volume). The first step is to establish the phylogenetic relationships of the host and parasite taxa independently of each other.

©The Systematics Association, 1986. This chapter is from *Coevolution and systematics* (eds A. R. Stone and D. L. Hawksworth) published for the Systematics Association by the Clarendon Press, Oxford.

The second step is to compare the phylogenies of the host and parasite groups; the assumption is that they should show congruence, and can therefore be used to establish 'true' relationships between taxa which otherwise would have remained unresolved. Although this approach is generally associated with Hennig's phylogenetic systematics, Hennig (1979) drew heavily on earlier work which had used parasite taxonomy to resolve relationships between animal host taxa. Quite independently of Hennig's work, investigators in other fields have invoked similar principles; for example, the use of parasite specialization in plant taxonomy (Meeuse 1973), and the classification of parasitic fungi (Raper 1968).

The assumption that host and parasite phylogenies are intimately intertwined is essential to the parasitological method and is based on assumptions about host-parasite biology, either implicitly or explicitly. The most recent comprehensive explicit statements of these assumptions have been given by Brooks (1979a), and Mitter and Brooks (1983):

Fahrenholz's Rule. Parasite phylogeny mirrors host phylogeny.

Szidat's Rule. The more primitive the host, the more primitive the parasites that it harbours.

Manter's Rules. (1) Parasites evolve more slowly than their hosts.

(2) The longer the association with a host group, the more pronounced the specificity exhibited by the parasite group.

(3) A host species harbours the largest number of parasites in the area where it has resided longest, so if the same or two related species of a host exhibit disjunct distribution and possess similar parasite floras, the areas in which the hosts occur must have been contiguous at a past time.

Eichler's Rule. The more genera of parasites a host harbours, the larger the systematic group to which the host belongs.

These rules have their origins predominantly in animal parasitology and have dominated the literature of the parasitological method. However, similar principles have been adduced in other fields. For example, Savile (1968) discussed methods for the classification of fungi and noted: 'In strict parasitism, the antiquity of the host reflects that of the parasite and *vice versa*' (cf. Szidat's Rule).

Johnson (1968) suggested that the specificity of parasites was a useful criterion in elucidating the relationships between parasitic fungi (cf. Fahrenholz's Rule). Indeed, in reviewing the literature of the systematics of plant host-parasite systems, it is rare to find any overlap with the literature of animal host-parasite systems.

The usefulness of the parasitological method depends critically on the relevant assumptions being satisfied in the host-parasite associations which are being examined, but the question remains of whether the assumptions are biologically realistic or whether they are 'rough guides, indicating the need for (and worth of) further analysis' (Ruse 1973). The importance of this question cannot be over-emphasized because it clearly discriminates between two approaches to the application of the parasitological method when the host and parasite phylogenies are *not* mirror images of each other. The first approach is to introduce *ad hoc* hypotheses to make the phylogenies fit each other and then seek to minimize the number of *ad hoc* assumptions required. The second approach interprets the lack of a good fit as evidence that the biology of the host–parasite relationships being investigated does *not* satisfy the assumptions of the model and, therefore, it is inappropriate to use the parasitological method.

Variation in virulence and resistance

Do individuals of a parasite species vary in their ability to attack host species and do individuals of a host species vary in their ability to resist infection by parasite species?

During the first quarter of this century, plant breeders had established the Mendelian basis of the inheritance of differences in the susceptibility of different varieties of crop species to various diseases (e.g. Biffen 1907). Plant pathologists had also established that individual isolates of some plant pathogens varied in their ability to infect different host varieties or lines and that this specificity was stable and inherited (e.g. Stakman and Levine 1919). In 1942 H. H. Flor, published the first of a series of papers in which he described experiments designed to establish the genetic basis of the *interaction* between flax (*Linum usitatissimum*) and flax rust (*Melampsora lini*), and continued in Flor (1946, 1956). Pairs of flax varieties which had opposite responses to two races of flax rust were crossed and at the same time the two rust races were crossed. The F1 and F2 generations of the host plant derived from these crosses were tested for resistance to the two races of the rust originally used to differentiate between the parental varieties. Conversely, the F1 and F2 generation derived from the rust crosses were tested for virulence (i.e. the ability to infect) on the original parental flax varieties. The results showed that resistance in the host lines was inherited as a single major gene with resistance dominant to susceptibility and that virulence was inherited as a single major gene recessive to avirulence (see Table 1.1). The consistency with which he obtained similar results in many crosses led him to propose that the interaction between flax and flax rust was controlled by complementary genetic systems in the host and parasite, and he summarized his interpretation as the 'gene-for-gene hypothesis':

'For every gene conditioning resistance in the host there is a corresponding gene conditioning virulence in the parasite.' (Flor 1956).

Table 1.1. Segregation of resistance (data from Flor 1947)

	Reaction of host		
Race of rust	Ottawa	Bombay	F1
21	S[a]	R[b]	R
22	R	S	R

F2: numbers of each class of reaction

	Reaction to race 21	
Reaction to race 22	R	S
R	110	32
S	43	9

Segregation of virulence (data from Flor 1946)

	Reaction of parasite		
Host variety	Race 21	Race 22	F1
Ottawa	V[c]	A[d]	A
Bombay	A	V	A

F2: numbers of each class of reaction

	Reaction on Ottawa	
Reaction on Bombay	A	V
A	78	27
V	23	5

(a) S = susceptible.
(b) R = resistant.
(c) V = virulent.
(d) A = avirulent.

Since then a number of different plant host–parasite interactions have been shown to be controlled by gene-for-gene systems, for example, powdery mildew (*Erysiphe graminis*) on cultivated barley (*Hordeum vulgare*; Moseman 1959; Day 1974; Vanderplank 1982). Although the gene-for-gene hypothesis conveniently describes the form of such interactions, there is indirect evidence that specificity exists at other levels, although its genetic basis is not understood. For example, Chin and Wolfe (1984) demonstrated that isolates of powdery mildew carrying virulence for the barley cultivars 'Hassan' and 'Wing', (which have different resistance genes), are therefore able to attack both, and show greater pathogenicity

Table 1.2. Virulence of isolates of *Erysiphe graminis* f.sp. *hordei*, obtained from different sources and able to attack both cv. 'Hassan' and cv. 'Wing'[a]

(A) Isolates obtained from pure stands of 'Hassan' or 'Wing' and tested on 'Hassan' and 'Wing' seedlings. Results given in terms of virulence ($\times 10^{-3}$ colonies/spore).

Source cultivar	Test cultivar		Mean
	Hassan	Wing	
Hassan	80.2[b]	65.6	72.9
Wing	55.6	68.6	62.7
Mean	68.4	67.2	

SED I[c] = ± 3.82; SED II[d] = ± 2.70.
Interaction (source cv. × test cv.) $P<0.01$.

(B) Isolates obtained from 'Hassan' or 'Wing' plants within mixed stand of 'Hassan', 'Midas' and 'Wing' and tested on 'Hassan' and 'Wing' seedlings. Results given in terms of virulence ($\times 10^{-3}$ colonies/spore).

Source cultivar	Test cultivar		Mean
	Hassan	Wing	
Hassan	56.6[b]	60.4	58.5
Wing	50.2	54.2	52.2
Mean	53.4	57.3	

SED I[c] = ± 5.02; SED II[d] = ± 3.50.
Interaction (source cv. × test cv.) not significant.
(a) Data from Chin and Wolfe (1984).
(b) Values in the body of the table are means of five isolates.
(c) SED I (8 df) for comparison of source cv.–test cv. combinations.
(d) SED II (8 df) for comparison of means of source or test cultivars.
(e) SED III (16 df) = ± 4.46 for comparisons between pure and mixed stands.

on the cultivar from which they were originally isolated (Table 1.2).

This admittedly cursory examination of the evidence does show that variation can exist in both host and parasite which could allow host populations to diverge in their ability to resist infection by pathogens and that variation exists in parasite populations which could allow parasites to track changes in host populations and lead to specialization on particular host genotypes. (For a more full discussion see Barrett 1983.)

On the other hand, there are many examples of plants and their parasites which do not exhibit this type of variation. Resistance to loose smut (*Ustilago nuda*) in some barley varieties has been achieved by selection for a cleistogamous (closed) flower habit which prevents entry of the fungal spores (Macer 1960). Resistance to southern corn leaf blight (*Cochliobolus heterostrophus*; anamorph *Drechslera maydis*) in maize has been progressively increased since the beginning of this century, but no new races of the parasite have been detected which can overcome this increased resistance (see Russell 1978). Similarly, variation in non-specific pathogenicity can exist in parasite species, for example *Ustilago hordei* (Emara and Sidhu 1974), and the Dutch elm disease fungus (*Ceratocystis ulmi*; Brasier 1979).

Host and parasite evolution

Can parasites affect the way in which a host population evolves and can parasites 'track' changes in host populations?

It is commonly observed that in intensively cultivated crops the introduction of a new variety resistant to a disease is followed by a period in which the variety remains free from disease and the area planted with the variety increases. After a number of years, 3–5 in the case of powdery mildew of barley (Fig.1.1), new genotypes of the pathogen which can overcome the resistance become so frequent in the parasite population that the variety falls from favour with the growers. Such processes are known as 'boom-and-bust' cycles. Although this is a rather artificial example, because the host response is dictated by man, it does show that the elements of coevolutionary processes can be reproduced experimentally.

There have been relatively few investigations of evolution in host–parasite systems which have not been subject to manipulation by man. Perhaps the best studies have been those of the interaction between the European rabbit (*Oryctolagus caniculus*) and myxoma virus (Fenner and Ratcliffe 1965; Fenner and Myers 1978; Mead-Briggs 1977). Because of the economic importance of rabbits in Australia, the effectiveness of myxoma virus as a control measure has been closely monitored; these studies provide probably the best and most comprehensive set of data on the dynamics of evolution in a host–parasite system. In the period which followed the introduction of myxoma virus into rabbit populations, there was a decrease in the virulence of the virus population and an increase in the resistance to the virus in the rabbit population (Tables 1.3 and 1.4). More recent data (Fenner and Myers 1978) suggest that there has been a slight increase in the virulence of the Australian myxoma population as a response to the decreased susceptibility of the rabbit population.

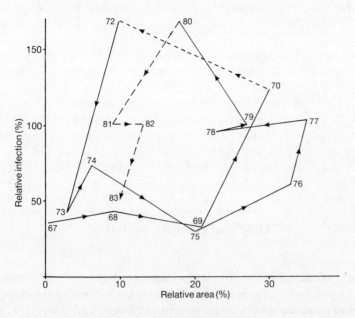

Fig.1.1. Relative infection of varieties carrying M1a 12 mildew resistance plotted against the relative area planted with these varieties in the UK between 1967 and 1983. Between 1973 and 1975 varieties carrying M1a 12 resistance were replaced by varieties carrying M1a 12 plus another resistance gene. [Data mostly from Agricultural Development Advisory Service, Harpenden Laboratory, *National Surveys of Foliar Diseases*, 1967–1980, and M. S. Wolfe (pers. comm. 1984).]

Table 1.3. Frequency of strains of myxoma with different levels of virulence[a]

	Grade of Severity					
	I	II	IIIA	IIIB	IV	V
Mean survival time (days)	13	14–16	17–22	23–28	29–50	—
Kill rate (%)	99	95–99	90–95	70–90	50–70	50
Australia						
1950–51	100	0	0	0	0	0
1958–59	0	25	29	27	14	5
United Kingdom						
1953	100	0	0	0	0	0
1962	4	18	39	25	14	1

(a) Data from Fenner and Ratcliffe (1965).

Table 1.4. Symptoms of myxomatosis in wild rabbits exposed to successive epizootics at Lake Urana in Australia[a]

Date	Number of epizootics	Symptoms		
		Severe to Fatal	Moderate	Mild
Wild rabbits before the introduction of myxoma	0	93	5	2
1953	2	95	5	0
1954	3	93	5	2
1955	4	61	26	13
1956	5	75	14	11
1958	7	54	16	30

(a) Data from Fenner and Ratcliffe (1965).

The myxoma-rabbit studies are unique in that both participants in the interaction have been studied over a long period, and this has allowed the evolutionary dynamics of the system to be observed. However, there are other studies from which the inferences can be drawn that parasites can produce genetic change in host populations. For example, wild rats captured from urban areas affected by epidemics of the plague bacillus *Yersinia pestis* in the recent past, show higher levels of resistance to infection by the bacillus than rats from areas with no recent records of plague (see Levin *et al.* 1982). Wahl and his coworkers have shown that in the highlands of Israel where conditions are not suitable for powdery mildew infection, barley plants show lower levels of resistance than plants from areas where powdery mildew develops more easily (e.g. see Wahl *et al.* 1978). Evidence of this type suggests that host populations can respond to the presence of parasites and that parasite populations can track changes in host populations. If the host populations also diverged to such an extent that the populations could be recognized as different species, then it seems reasonable that one would find specialization to each of these species, although the precise taxonomic status of the parasite populations cannot be predicted easily.

On the other hand, there are many examples of changes in host populations which have not been accompanied by corresponding changes in the parasite populations. Despite the widespread use of the cleistogamous flower habit in cultivated barley in western Europe, there has been no apparent response from the smut fungus. The Gros Michel clone of banana formed the basis of the Caribbean and Central American banana trades but was devastated by Panama disease (caused by the fungus *Fusarium oxysporum* f.sp. *cubense*) during the earlier part of this century.

The Cavendish clone and its derivatives were used to replace Gros Michel because they showed resistance to this disease. Despite many years of intensive cultivation, there has been no evolution of fungus strains specific for the Cavendish clone (Simmonds 1966; Wilhelm 1981).

Thus, although some parasite populations can track changes in host populations and, indeed, cause genetic divergence between host populations, these processes do not necessarily occur. Conversely, it is possible for hosts to change in respect of their interaction with their parasites with the parasites failing to follow.

Parasite specificity and host phylogeny

Do parasite species show specificity for host species or groups of species that share a common phylogeny?

Powdery mildew (*Erysiphe graminis*) can be found on cultivated wheat, barley, oats, and rye. If isolates are taken from each host species and used to reinfect all of them, it is generally found that each isolate grows best on the species from which it was originally obtained. *Erysiphe graminis* can, therefore, be subdivided into groups reflecting the host species specificity of each isolate, termed special forms (Jenkyn and Bainbridge 1978), viz:

Barley: *E. graminis* f.sp. *hordei*
Wheat: *E. graminis* f.sp. *tritici*
Oats: *E. graminis* f.sp. *avenae*
Rye: *E. graminis* f.sp. *secalis*

This sort of specificity can be found in many of the economically important diseases of crop plants, for example, *Puccinia graminis* (stem rust; Anikster 1984). Indeed, a few plant pathologists split fungal species into special forms whenever the fungus can be found on more than one host species, even in the absence of any study of cross-infection. Where cross-infection studies have been undertaken and the phenotypic basis of the specificity is reasonably well established, it is of interest to know the genetic basis of the specialization.

A number of studies examined the host specificities of hybrids between different special forms of plant pathogenic fungi (e.g. Sanghi and Luig 1974; Green 1971; see also reviews by Anikster 1984; Hiura 1978; Scott *et al.* 1979). The hybrids and generations derived from them showed a number of different effects. Some showed reduced pathogenicity for one or both of the species attacked by the parental strains, and some showed wider host ranges than either of the parents. In some cases reciprocal crosses differed in fertility. Although the genetic basis of specialization is not known in detail (cf. gene-for-gene hypothesis), it is obvious that sufficient genetic variation exists in pathogen populations to produce substantial changes in host specificity and that a division into special forms is often somewhat artificial. It has also been demonstrated that even within a single host species, variation with respect to resistance to special forms from other host species may exist. For example, Sanghi and Baker (1972) were able to select a wheat line susceptible to the majority of strains of *Puccinia graminis* f.sp. *avenae* used in their experiments.

Newton *et al.* (1985) have recently shown that although yellow rust of both wheat and barley contains many different races, the races specializing on each host species are identical with respect to 12 electrophoretic markers, but the barley attacking races differed at two of these markers from those attacking wheat. Again using electrophoresis, Burdon *et al.* (1981) demonstrated that a novel special form of stem rust able to attack the wild grass *Agropyron scabrum* arose from hybridization between special forms on wheat and rye. The sexual stage of *P. graminis* is not found in Australia and so the hybridization must have occurred by somatic recombination.

In 1946 a new cultivar of oats, 'Victoria', was planted on the northern plains of the USA. Within 2 years it had been devastated by a previously unknown disease of oats caused by a fungus, *Cochliobolus victoriae* (anamorph: *Drechslera victoriae*; Day 1974). This fungus was later identified as a minor parasite on wild grasses in the area. Other cultivated varieties of oats, unrelated to cv. 'Victoria', were not susceptible to the disease. The genetic basis of the susceptibility of 'Victoria' has been shown to be relatively simple and due either to a pleiotropic effect of the gene for resistance to crown rust (*P. coronata*) carried by cv. 'Victoria' or a gene closely linked to this resistance gene.

Two conclusions follow from these examples. Parasite species may shift host specificity fairly easily and the genetic changes required may be fairly simple. Where host species show resistance to parasites of other species, this may be controlled by relatively few genes. It is fairly easy to envisage a situation in which parasite species are capable of attacking a wide range of different host species, but the degree of specialization for a particular host species, and hence the ease with which special forms can be recognized, is dependent on the species composition and structure of the ecological community and the genetic composition of the various host species.

The examples quoted above have been mainly of parasite species able to attack a range of different host species but which show some specificity for each host species, but how representative are they? It is possible to find parasite species that are confined to a single host species, such as *Ustilago avenae* on oats. There are yet other parasite species which are able to attack a range of different host species without any apparent specialization on each, for example, the ergot fungus (*Claviceps purpurea*) on cultivated cereals (Mantle *et al.* 1977) and *P. violae* which appears to only infect members of the genus *Viola* (Wheeler 1968). In these cases there is some taxonomic consistency in the host range but there are parasite species which are quite promiscuous in their host preferences. Panama disease of banana is caused by *Fusarium oxysporum*, but the race which attacks banana, is just one of a range of over 20 special forms with different host specificities (Nelson 1981; Snyder and Hansen 1940; see also Parlevliet, this volume), for example:

Banana: *F. oxysporum* f.sp. *cubense*
Flax: *F. oxysporum* f.sp. *lini*
Tomato: *F. oxysporum* f.sp. *lycopersici*
Pea: *F. oxysporum* f.sp. *pisi*
Cotton: *F. oxysporum* f.sp. *vasinfectum*

Apart from these hosts all being vascular plants there is little taxonomic consistency in the host range of *F. oxysporum*.

Discussion of host–parasite interactions generally centres around the assumption that the relationship is essentially antagonistic, but there are situations in which the maintenance of a parasite population within a host population can be viewed as a defence mechanism of the host preventing immigration of more susceptible competitors. Barbehenn (1969) has argued that just such a mechanism could account for the high mortality in moose whenever they come into contact with populations of white-tailed deer and become infected with the meningeal worm *Parelaphostrongylus tenuis*. Such mechanisms could have profound effects on evolution in the host species. Aggressive exploitation of parasite faunas could split a species into reproductively isolated groups with subsequent effects on phylogenesis. The involvement of parasites in preventing genetic introgression between populations can, perhaps be seen most clearly in the *Culex pipiens* complex. This mosquito group consists of a number of morphologically almost identical populations, between which it is almost impossible to obtain fertile matings. A major factor in this genetic isolation is that each population carries a different rickettsia-like cytoplasmic parasite (*Wolbachia pipientis*) which is involved in the incompatibility mechanism in a way not yet fully understood. That the parasite fulfils this role can be demonstrated by treating the mosquitoes with an antibiotic; after treatment fertile matings between individuals from incompatible populations can be obtained (Yen and Barr 1974).

There is little doubt that parasite species can be found which show specificity for single host species or groups of related host species, but they are by no means typical. Parasite species with very wide host ranges can just as easily be found.

Parasite specialization in natural ecosystems

How specialized are 'specialized' parasites in natural ecosystems?

Many of the examples so far cited involve interactions between a parasite and a cultivated host plant species. It is always possible to argue that agriculture is a special case and that it is not possible to extrapolate very easily from it to natural ecosystems (Barrett 1985). The study of plant pathogens is almost entirely an applied science and so there have been relatively few investigations of plant host–parasite interactions under natural conditions. The following examples are of plant parasites, which under agricultural conditions show strong host specificity. Isolates of *P. coronata* and *E. graminis* have been obtained from wild grass species in natural ecosystems in Israel and used to infect a range of potential host species (e.g. Eshed and Dinoor 1981, Wahl *et al.* 1978). The data (Table 1.5) show that these isolates had wide host-ranges. What is not so obvious from these data, is that each isolate did not infect all members of each genus or tribe, but several from each; for an example see Table 1.6. In another study, Gerechter-Amitai (1973) showed that the host ranges of f.sp. *tritici* and f.sp. *avenae* of *Puccinia graminis* were very wide under laboratory conditions and that even under field conditions the host ranges were still spread over several genera and tribes, although this was less than in the glasshouse (Table 1.7) (see also Wahl *et al.* 1984).

Table 1.5. Host ranges of fungal parasites isolated from wild grasses in Israel

Source Species	Number of species infected	Number tested		
		species	genera	tribes
Erysiphe graminis[a]				
Hordeum murinum	17	44	13	4
Bromus rigens	18	30	11	3
Phalaris paradoxa	33	40	20	6
Alopecurus myosuroides	53	74	30	6
Puccinia coronata[b]				
Agrostis verticillata	69	106	37	7
Arrhenatherum palaestinum	13	106	12	2

(a) Data from Wahl *et al.* (1978).
(b) Data from Eshed and Dinoor (1980, 1981).

It would appear from these observations that the level of specificity shown by a parasite can depend on the ecological opportunities available to it. Under uniform agricultural conditions, these parasites tend towards host-species specificity, but under the more heterogeneous conditions in natural ecosystems, they tend towards wider host ranges.

Table 1.6. Host range of an isolate of *Erysiphe graminis* from *Hordeum murinum*[a]

Tribe	Genus	Number of species tested	Number of species successfully infected
Hordeae	*Elymus*	2	1
	Eremopyrum	1	1
	Hordeum	4	3
	Psilurus	1	1
Festuceae	*Bromus*	15	3
	Cutandia	3	1
	Echinaria	1	1
	Lamarckia	1	1
	Lolium	5	1
	Sphenopus	1	1
	Vulpia	5	1
Avenae	*Avena*	4	1
Stipeae	*Stipa*	1	1

(a) Data from Wahl *et al.* (1978).

Table 1.7. Host range of *P. graminis* on wild grasses from Israel under different experimental conditions[a]

Special forms	Greenhouse		Field	
	Number of species infected	Number of genera infected	Number of species infected	Number of genera infected
f.sp. *avenae*	82	41	66	34
f.sp. *tritici*	49	20	27	8

(a) Data from Gerechter-Amitai (1973), quoted by Browning (1980).

Discussion

The 'rules' which underpin the parasitological method have not been examined individually in this paper, but the examples that have been used should indicate that none of them can be taken as universally 'true'. As far as a parasite is concerned, its host is just part, albeit a major part, of its ecological niche. Viewed from this perspective why should parasites not show the same range of ecological adaptations and life-styles adopted by other organisms not involved in intimate associations? Why should there not be specialist/generalist, pioneer/climax parasite species? Most of the examples used to illustrate this contribution have been drawn from plant host-parasite interactions, but similar ranges of interactions can be found among animal host-parasite interactions (for example, Holmes 1983):

> 'Animal-helminth systems appear to be evolving along a variety of pathways, some leading to greater host specificity, some to less, some leading to high levels of pathogenicity, some to low pathogenicity or even to mutualism.'

It is undoubtedly true that cases can and have been analysed in which host and parasite species are associated by descent and therefore amenable to analysis using the parasitological method (e.g. Brooks 1979*b*). However, the complex interplay of ecological forces in the biology of host-parasite interactions is such that association by descent is only one of the possible outcomes. One is left with the rather unsatisfying conclusion that the parasitological method, *sensu stricto*, will work where it works.

References

Anikster, Y. (1984). The formae speciales. In *The Cereal Rusts*. Vol. 1. *Origins, Specificity, Structure and Physiology* (eds W. R. Bushnell and A. P. Roelfs), pp. 115-30. Academic Press, New York.

Barbehenn, K. R. (1969). Host-parasite relationships and species diversity in mammals: an hypothesis. *Biotropica* 1, 29-35.

Barrett, J. A. (1983). Plant-fungus symbioses. In *Coevolution* (eds D. J. Futuyma and M. Slatkin), pp. 137-60. Sinauer Associates Inc., Sunderland, Massachusetts.

—— (1985). The gene-for-gene hypothesis: parable or paradigm. In *Ecology and genetics of host-parasite interactions* (eds D. Rollinson and R. M. Anderson). Linnean Society, Symposium Series No. 11, pp. 215-25. Academic Press, London.

Biffen, R. H. (1907). Studies in the inheritance of disease resistance. *J. Agric. Sci.* 2, 109-28.

Brasier, C. M. (1979). Dual origin of recent Dutch elm disease outbreaks in Europe. *Nature, Lond.* 281, 78-80.

Brooks, D. R. (1979*a*). Testing the context and extent of host-parasite coevolution. *Syst. Zool.* 28, 299-307.

—— (1979*b*). Testing hypotheses of evolutionary relationships among parasites: the digeneans of crocodilians. *Am. Zool.* 19, 1225-38.

Browning, J. A. (1980). Genetic protective mechanisms of plant-pathogen populations: their coevolution and use in breeding for resistance. In *Biology and Breeding for Resistance to Arthropods and Pathogens of Cultivated Crops* (ed. M. K. Harris), pp. 52-75. Texas A&M University Press, College Station, Texas.

Burdon, J. J., Marshall, D. R., and Luig, N. H. (1981). Isozyme analysis indicates that a virulent cereal rust pathogen is a somatic hybrid. *Nature, Lond.* 293, 565-6.

Chin, K. M. and Wolfe, M. S. (1984). Selection on *Erysiphe graminis* in pure and mixed stands of barley. *Plant Pathology* 33, 535-45.

Day, P. R. (1974). *Genetics of Host-Parasite Interaction*. W. H. Freeman & Co., San Francisco.

Emara, Y. A. and Sidhu, G. (1974). Polygenic inheritance of aggressiveness in *Ustilago hordei*. *Heredity* 32, 219-24.

Eshed, N. and Dinoor, A. (1980). Genetics of pathogenicity in *Puccinia coronata*: specialization at the host genus level. *Phytopathology* 70, 1042-6.

—— —— (1981). Genetics of pathogenicity in *Puccinia coronata*: the host range among grasses. *Phytopathology* 71, 12-9.

Fenner, F. and Myers, K. (1978). Myxoma virus and myxomatosis in retrospect: the first quarter century of a new disease. In *Viruses and environment* (eds E. Kurstak and K. Maramorosch), pp. 485-501. Academic Press, New York.

—— and Ratcliffe, F. N. (1965). *Myxomatosis*, Cambridge University Press, Cambridge.

Flor, H. H. (1946). Genetics of pathogenicity in *Melampsora lini*. *J. Agric. Res.* 73, 335-57.

—— (1947). Inheritance of reaction to rust in flax. *J. Agric. Res.* 74, 241-62.

—— (1956). The complementary genic systems in flax and flax rust. *Adv. Genet.* 8, 29-54.

Gerechter-Amitai, Z. K. (1973). *Stem rust, Puccinia graminis Pers., on cultivated and wild grasses in Israel*. PhD thesis, Hebrew University, Jerusalem. [Quoted in Wahl *et al.* (1984) and Browning (1980).]

Green, G. J. (1971). Hybridization between *Puccinia graminis tritici* and *Puccinia graminis secalis* and its evolutionary implications. *Can. J. Bot.* 49, 2089-95.

Hennig, W. (1979). *Phylogenetic Systematics*. (Translated by D. Dwight Davis and Rainer Zangerl.) University of Illinois Press, Urbana, Illinois.

Hiura, U. (1978). Genetic basis of formae speciales in *Erysiphe graminis* DC. In *The Powdery Mildews* (ed. D. M. Spencer), pp. 101-28. Academic Press, London.

Holmes, J. C. (1983). Evolutionary relationships between parasitic helminths and their hosts. In *Coevolution* (eds D. J. Futuyma and M. Slatkin), pp. 161-85. Sinauer Associates, Sunderland, Massachusetts.

Jenkyn, J. F. and Bainbridge, A. (1978). Biology and pathology of cereal powdery mildews. In *The Powdery Mildews* (ed. D. M. Spencer), pp. 283-321. Academic Press, London.

Johnson, T. (1968). Host specialization as a taxonomic criterion. In *The Fungi: an Advanced Treatise*. Vol. III. *The Fungal Population* (eds G. C. Ainsworth and A. S. Sussman), pp. 543-56. Academic Press, New York.

Levin, B. R., Allison, A. C., Bremermann, H. J., Cavalli-Sforza, L. L., Clarke, B. C., Frentzel-Beyme, R., Hamilton, W. D., Levin, S. A., May, R. M., and Thieme, H. R. (1982). Evolution of parasites and hosts. Group report. In *Population Biology of Infectious Diseases*, Dahlem Konferenzen 1982 (eds R. M. Anderson and R. M. May), pp. 213-43, Springer-Verlag, Berlin.

Macer, R. C. F. (1960). In *Annual Report of the Plant Breeding Institute, Cambridge*, p. 62.

Mantle, P. G., Shaw, S., and Doling, D. A. (1977). Role of weed grasses in the etiology of ergot disease in wheat. *Ann. appl. Biol.* 86, 339-51.

Mead-Briggs, A. R. (1977). The European rabbit, the European rabbit flea and myxomatosis. *Applied Biology* (ed. T. H. Coaker), vol. 2, pp. 183–261. Academic Press, London.

Meeuse, A. D. J. (1973). Coevolution of plant hosts and their parasites as a taxonomic tool. In *Taxonomy and Ecology* (ed. V. H. Heywood), pp. 289–316. Academic Press, London.

Mitter, C. and Brooks, D. R. (1983). Phylogenetic aspects of coevolution. In *Coevolution* (eds D. J. Futuyma and M. Slatkin), pp. 65–98. Sinauer Associates, Sunderland, Massachusetts.

Moseman, J. G. (1959). Host–parasite interaction of the genes for resistance in *Hordeum vulgare* and for pathogenicity in *Erysiphe graminis* f.sp. *hordei*. *Phytopathology* 219, 469–72.

Nelson, P. E. (1981). Life cycle and epidemiology of *Fusarium oxysporum*. In *Fungal Wilt Diseases of Plants* (eds M. E. Mace, A. A. Bell, C. M. Beckman), pp. 51–80. Academic Press, New York.

Newton, A. C., Caten, C. E., and Johnson, R. (1985). Variation for isozymes and double stranded RNA among isolates of *Puccinia striiformis* and two other cereal rusts. *Plant Pathology* 34, 235–47.

Raper, J. R. (1968). On the evolution of fungi. In *The Fungi: an Advanced Treatise*, Vol. III. *The Fungal Population* (eds G. C. Ainsworth and A. S. Sussman), pp. 677–93. Academic Press, New York.

Ruse, M. (1973). *The Philosophy of Biology*. Hutchinson University Library, Hutchinson & Co., London.

Russell, G. E. (1978). *Plant Breeding for Pest and Disease Resistance*. Butterworths, London.

Sanghi, A. K. and Baker, E. P. (1972). Genetic basis for resistance in two common wheat cultivars to stem rust strains of unusual virulence. *Proc. Linn. Soc. N.S.W.* 97, 56–71.

—— and Luig, N. H. (1974). Resistance in three common wheat cultivars to *Puccinia graminis*. *Euphytica* 23, 273–80.

Savile, D. B. O. (1968). Possible interrelationships between fungal groups. In *The Fungi: an Advanced Treatise*. Vol. III. *The Fungal Population* (eds G. C. Ainsworth and A. S. Sussman), pp. 649–75. Academic Press, New York.

Scott, P. R., Johnson, R., Wolfe, M. S., Lowe, H. J. B., and Bennett, F. G. A. (1979). Host-specificity in cereal parasites in relation to their control. *Plant Breeding Institute Annual Report*, pp. 27–62. Cambridge. Reprinted in: *Applied Biology* (ed. T. H. Coaker), vol V pp. 350–93. Academic Press, London.

Simmonds, N. W. (1966). *Bananas* (2nd ed). Longman, London.

Snyder, W. C. and Hansen, H. N. (1940). The species concept in *Fusarium*. *Am. J. Bot.* 27, 64–7.

Stakman, E. C. and Levine, M. N. (1919). New biologic forms of *Puccinia graminis*. *J. Agric. Res.* 16, 103–5.

Vanderplank, J. E. (1982). *Host–pathogen Interactions in Plant Disease*. Academic Press, New York.

Wahl, I., Anikster, Y., Maniterski, J., and Segal, A. (1948). Evolution at the center of origin. In *The Cereal Rusts*. Vol. 1. *Origins, Specificity, Structure and Physiology* (eds W. R. Bushnell and A. P. Roelfs), pp. 39–77. Academic Press, New York.

—— Eshed, N., Segal, A., and Sobel, Z. (1978). Significance of wild relatives of small grains and other wild grasses in cereal powdery mildews. In *The Powdery Mildews* (ed. D. M. Spencer), pp. 83–100. Academic Press, London.

Wheeler, B. E. J. (1968). Fungal parasites of plants. In *The Fungi: an Advanced Treatise*. Vol. III. *The Fungal Population* (eds G. C. Ainsworth and A. S. Sussman), pp. 179–210. Academic Press, New York.

Wilhelm, S. (1981). Sources and genetics of host resistance in field and fruit crops. In *Fungal Wilt Diseases of Plants* (eds M. E. Mace, A. A. Bell, and C. H. Beckman), pp. 299–376. Academic Press, New York.

Wolfe, M. S. (1984). Trying to understand and control powdery mildew. *Plant Pathology* 33, 451–66.

Yen, J. H. and Barr, A. R. (1974). Incompatibility in *Culex pipiens*. In *The Use of Genetics in Insect Control* (eds R. Pal and M. J. Whitten), pp. 97–118. Elsevier/North Holland, Amsterdam.

2. Coevolution of host resistance and pathogen virulence: possible implications for taxonomy

J. E. PARLEVLIET

Department of Plant Breeding, Agricultural University, Wageningen, The Netherlands

---

Abstract

Organisms evolve in relation to their environments of which, for parasites the hosts form an essential part, while for the hosts the parasites are of considerable importance. The adaptation of a parasite to exploit its hosts is not yet coevolution; it is an evolution of the parasite to fit its ecological niche. Coevolution is mutual evolution: the host evolving defence mechanisms, the parasite evolving mechanisms to circumvent or neutralize these defences. When different hosts are parasitized by a group of parasites the hosts evolve varying defence mechanisms. A parasite adapting to a defence mechanism adapts to the host species employing that defence system: it specializes. Coevolution therefore tends towards a higher degree of specialization among parasites in terms of host range. Among pathogens, parasites of fungal, bacterial, or viral origin, high degrees of specialization have developed. The host range of fungal pathogens is often restricted to some species of one host genus only such as *Erysiphe trina* on *Quercus*, *E. sordida* on *Plantago*, and *Puccinia hordei* on *Hordeum* (and on *Ornithogalum* as alternate host). In such coevolved systems the genes for pathogenicity or virulence in the pathogen and the genes for resistance in the host have evolved in a gene-for-gene way. This indicates a highly intricate relationship possibly of great value in the study of the taxonomic relationships of the host taxa. This latter conclusion, however, should be approached with caution. There are

©The Systematics Association, 1986. This chapter is from *Coevolution and Systematics* (eds A. R. Stone and D. L. Hawksworth) published for the Systematics Association by the Clarendon Press, Oxford.

indications that pathogens may switch from one host to another more easily than this intricate relationship suggests.

Introduction

Organisms in natural ecosystems can be classified as *producers* (green plants), *consumers* satisfying their food requirements from producers, other consumers or reducers, and *reducers*. Parasites are consumers that live in close association with their hosts. Parasites of a non-animal nature such as viruses, bacteria, and fungi are often termed pathogens.

Both hosts and pathogens evolve to fit the ecological niche they occupy. The evolution of the host is rarely determined to any significant degree by the evolution of a single pathogen. The evolution of the pathogen, however, especially when highly specialized, can be expected to be determined to a much greater extent by the evolution of its host as the host constitutes such an important part of its ecology. Adaptation of pathogens to the seasonal growth patterns and the morphological, anatomical, and physiological features of the hosts in order to survive is unavoidable. Such adaptations lead to specialization amongst pathogens even without adaptation to special defence mechanisms in the hosts. This adaptation to the host, however complicated and precise it may be, is not yet coevolution.

Integrated into these evolutionary developments coevolution between host and pathogen may occur when host and pathogen evolve an interdependence with each other. This is the concept of coevolution advocated by Dobzkansky (1970), and Ehrlich and Raven (1964). However, a second meaning is used by some authors, including Anikster and Wahl (1979), and Savile (1971), who describe the evolution of the parasite in relation to the evolution of the host; i.e. the host plants evolve and the pathogens coevolve with them.

Coevolution in the former sense is the interpretation adopted here and an excellent example is that of leguminous host plants and the nitrogen-fixing bacteria of the genus *Rhizobium* which coevolved to form a mutualistic symbiosis.

Coevolution of host and pathogen: a model

Imagine a number of host species are all equally parasitized by several unspecialized pathogen species (Parlevliet 1979). All the host and pathogen species are genetically heterogeneous, individuals within a host species may therefore vary in the ease with which they can be exploited by the pathogens. Those that are less easy to exploit have a slight reproductive advantage and the characteristics on which this is based tend to become selected. The defence mechanism that starts to evolve in this phase is most likely to be of a general nature, effective against all pathogens, as such defence mechanisms have the highest survival value. The pathogens are gradually confronted with host species increasingly difficult to exploit, and moreover they are faced with varying defence mechanisms developed independently from each other and originating from different gene pools.

Within the pathogen, species differences may exist in the capacity to use host species employing a certain defence mechanism. Genotypes which can exploit host species with that defence mechanism tend to become selected. The pathogen adapts to that defence mechanism and so it adapts to the host species carrying that defence mechanism. Other pathogens may develop other adaptations to this or to other defence mechanisms, ultimately leading to a divergence between host species in terms of defence mechanisms and amongst pathogens adapted to the various defence mechanisms and so to the various host species exploiting these mechanisms.

This I consider is coevolution. The first defences of the host are of a general nature, such as phytoalexins (low molecular weight anti-microbial compounds produced by plants in response to attacks by pathogens), effective against whole groups of pathogens, but when pathogens become specialized new defence mechanisms are directed towards a narrower range of pathogens. Ultimately, a situation arises where the host species employ diverse forms of general defence mechanisms and various forms of specialized defence mechanisms. The pathogens that have survived must therefore have adaptations to some defences, the set of defence mechanisms adapted to varying from one pathogen to the other; a complex system of host–pathogen relationships has evolved.

Specialization

A parasitic way of life itself is already a specialized one, but the specialization goes considerably further. The method of parasitism, the parts of the host parasitized, the entering of the host, the way of bridging non-host periods, the reproduction, spread, and host range are all subject to adaptations.

Viruses are intracellular pathogens, while many fungi feed intercellularly through haustoria or from intercellular spaces. Some fungi are even ectoparasites, remaining on the surface and penetrating the host cells only with their haustoria (e.g. powdery mildews). Some fungi parasitize the living host cells (biotrophs), others kill the host tissue to be parasitized (necrotrophs).

Pathogens may also specialize on certain parts of the host. Many viruses are systemic, to be found in the whole host. Fungi on the other hand may be very specific with respect to the tissue parasitized. Some pathogens parasitize only the aerial parts of the host plants like the various rust species of the cereals (leaf rust, *Puccinia hordei*, on barley; stem rust, *P. graminis*, of wheat, barley, rye, and oats). Others mainly attack the reproductive organs (such as loose smut, *Ustilago nuda* of wheat and barley) or roots (take all, *Gaeumannomyces graminis*, of cereals).

The way pathogens survive and/or are dispersed is often shown in the way it is referred to by pathologists; vector-borne (many viruses), seed-borne (some viruses, some fungi), air-borne (many fungi like the powdery mildews and the cereal rusts), or soil-borne (many fungi, like the take-all of the grasses).

The type of specialization that may have implications for coevolution and for the taxonomist is specialization in host species, i.e. the host range.

Host ranges may vary widely. Some pathogenic species have a broad range of hosts covering many families in different orders (generalists), others are restricted to genera in one family, and others are only found on a few species in one genus or even single species (specialists). The soil-borne fungus *Sclerotinia sclerotiorum*, is a typical generalist; it parasitizes hundreds of plant species from at least 64 families belonging to both the dicotyledons and the monocotyledons. *Erysiphe graminis* is a powdery mildew of grasses, Poaceae, whereas *E. sordida* and *E. trinia* are typical specialists, the latter being restricted to species of the genera *Plantago* and *Quercus*, respectively. *Alternaria kikuchiana* and *A. mali* are examples of extreme specialization, they are restricted to a few cultivars of the Japanese pear (*Pyrus serotina*) and apple (*Malus pumila*), respectively.

This specialization in host range was called biological specialization by Leppik (1965), and appears to occur especially among biotrophic fungi such as powdery mildews, rusts, and smuts. Such specialization, though, may follow unexpected pathways varying greatly from restriction to only one species, to a wide range of species of one family, or even covering several families. Pathogens classified in one genus tend to have their hosts also in taxonomic clusters (Hijwegen 1979), but exceptions such as *Erysiphe*, found on monocotyledons and dicotyledons, are not particularly rare. In the case of pathogens with host alternations, such as many rusts, the degree of specialization can vary with the phase. The blister rust, *Cronartium asclepiadeum*, occurs on a few species of *Pinus* sect. *Pinus* as the aecial stage, whereas the telial stage occurs on 34 species from 10 dicotyledonous families (Leppik 1967).

In these examples specialization at species level of the parasite is involved, but specialization may operate within the species. The powdery mildew of grasses is a serious leaf pathogen of wheat, barley, rye, and oats, but the mildew found on each appears avirulent on the others.

Erysiphe graminis is consequently recognized as having several special forms, viz. f.sp. *tritici*, *hordei*, *secalis*, and *avenae*. In the same way the stem rust, *Puccinia graminis*, the leaf rust, *P. recondita*, and the crown rust, *P. coronata* can be subdivided into special forms (Anikster and Wahl 1979). This specialization is often said to be more pronounced on the cultivated crops than in natural habitats (Eshed and Dinoor 1980), but has scarcely been searched for on native plants.

Within special forms, races are found extensively in many crop-pathosystems. In *Erysiphe* and *Puccinia* large numbers of races have been identified, each race being adapted to one or more host cultivars or rather adapted to (virulent against) certain resistance genes carried by those cultivars. A race therefore can be defined as the collective of all pathogen genotypes carrying the same combination of virulence genes (Parlevliet 1985). As in modern farming cultivars carrying certain resistance genes can be grown over very large acreages, races virulent on such cultivars may and do develop. In the natural host–pathogen situation, these pathogen genotypes occur not as a local race, but as one of the genotypic components of a genetically diverse population.

Biological specialization and coevolution

Specialization is a characteristic feature of pathogens, but the procedure and the degree of host specialization vary considerably. Host specialization may be part of the process of coevolution between pathogen and host, but can be also just a process of specialization to the host without any reciprocal adaptations involved.

As defined in this paper, coevolution almost unavoidably implies a certain degree of host specialization but the reverse is not necessarily true; that depends on the type of pathogen. For biotrophic pathogens such as powdery mildews, rusts, and smuts, this may very well be true, but there are large groups of pathogens with a quite different way of parasitic life where specialization is not an indication of coevolution. The toxin-producing pathogens, for instance, are often quite specialized, the host specialization being entirely or partly dependent on specific toxins (Scheffer and Livingston 1984; Mitchell 1984). *Cochliobolus victoriae* and *C. carbonus* each produce a different toxin and also have different hosts, oats, and maize, respectively. These two fungal species can be crossed and their progeny segregates into four groups of equal size; those that can attack both oats and maize, oats alone, maize alone, and neither crop. The capacity to attack oats and/or maize is entirely dependent on the production of the host-specific toxin, the toxin production being controlled by single genes (Comstock and Scheffer 1973; Scheffer and Livingston 1984). Recombination of such toxin-producing genes can easily lead to changes in host range; specialization then is a relative concept.

An excellent example of the versatility of toxin-producing pathogens is provided by the *Alternaria alternata* complex. *A. alternata* is primarily a saprophyte or weak pathogen attacking only senescent, weakened, or damaged tissue (Nishimura *et al.* 1979). To attack healthy plant tissue such species need toxins and several host crops are affected seriously by host-specific toxin producing *Alternaria* species (Table 2.1). These taxa are essentially classified on the basis of their hosts and therefore the toxins they produce (Yoder 1980). They are identical to the saprophytic or weakly pathogenic races that occur universally and which are included in *Alternaria alternata* by Nishimura *et al.* (1979). The fungi of this complex exhibit a general aggressiveness on the dead or senescent tissue of a very broad range of host species, independent of host-specific toxin production. From this large *Alternaria* gene pool races also arise that produce specific toxins by which they can vigorously attack certain hosts. The toxins are extremely host-specific, effective against only one or a few cultivars of one host species (Table 2.1), and may be diverse in structure as in the case of the AL and AK toxins (Scheffer and Livingston 1984).

Table 2.1. Some *Alternaria* taxa, their toxins, diseases they cause and their hosts (after Nishimura *et al.* 1979; Scheffer and Livingston 1984)

Pathogen and toxin	Toxin designation	Disease	Susceptible host
A. alternata f.sp.			
lycopersici toxin	AL	Stem canker	Tomato, cv. Earlypak 7
A. citri toxin			
(tangerine race)	ACT	Brown spot	Dancy tangerine
A. citri toxin			
(lemon race)	ACL		Rough lemon
A. kikuchiana toxin	AK	Black spot	Japanese pear, some cvs.
A. mali toxin	AM	Alternaria blotch	Apple, some cvs.

Alternaria in citrus is an example. It occurs as a weak or benign pathogen on senile tissues of various citrus crops, but virulent forms became locally destructive in Australia in 1964 on the 'Emperor Mandarin' and in 1974 in Florida on the closely related 'Dancy Tangerine'; in Florida another form occurs on the rough lemon. These forms are morphologically identical with the weakly pathogenic, non-specialized *Alternaria*, but differ in their capacity to produce host-specific toxins (Scheffer and Livingston 1984). The Dancy Tangerine toxin seems identical or at least very similar to the Emperor Mandarin toxin; they have the same host specificity (the ACT toxin). The rough lemon form, however, produces another toxin, the ACL toxin (Kohmoto *et al.* 1979).

Coevolution of resistance and virulence

This specialization has little to do with coevolution. Yoder (1980) suggests that in both *Alternaria* and *Cochliobolus* there is a germplasm pool which has a basic aggressiveness and from this pool forms with the capacity to produce toxins (apparently not too difficult to acquire) emerge, that determine the pronounced host specificity.

Fusarium oxysporum is another example where specialization occurred, probably without coevolution of any significance being involved. *Fusarium oxysporum* is a half-parasite and can live saprophytically in the soil for long periods (Kiraly *et al.* 1970). In nature it may act as a generalist, but in relation to the crops grown by man it has specialized remarkably. Gordon (1965) already recognized 66 special forms, each being able to attack only one or a few related crop species (Table 2.2). In a few cases the special forms are less clearly delineated (Armstrong and Armstrong 1975). The f.sp. *pisi* can only attack peas, but the f.sp. *apii* has a fairly wide host range and may also occur on peas; a similar situation to that in *Alternaria* may exist in this case. *Fusarium oxysporum* seems to exist as a large unspecialized germ plasm pool, from which races have developed which are specialized to certain crop species. Such specialization is of considerable advantage to the fungus in the case of exposure to monocultures of crop species.

Table 2.2. Some of the many special forms of *Fusarium oxysporum* and its host crop species and the identified races within the formae

Special forms	Host	Races
carthami	Safflower, *Carthamus tinctorius*	4
dianthi	Carnation, *Dianthus caryophyllus*	1
cucumerinum	Cucumber, *Cucumis sativus*	1
melonis	Musk melon, *Cucumis melo*	4
lycopersici	Tomato, *Lycopersicon esculentum*	3
pisi	Pea, *Pisum sativum*	several

The pathogenic bacterium *Xanthomonas campestris* is a similar case with over 120 pathovars recognized. Each pathovar is restricted to one genus or at most a few related genera (Leyns *et al.* 1984).

The above mentioned cases appear to represent a type of specialization consisting of a branching out of highly specialized races from a population that can be classified as a generalist. This specialization does not initially involve coevolution, but once the more specialized relationship is established the mutual evolution of resistance and virulence may develop. *Alternaria alternata*, *F. oxysporum*, and *Xanthomonas oryzae* may represent various stages in the evolution of this type of specialization from the least to the most advanced. The bacterial example seems to represent an advanced case as many of the pathovars were first classified as species.

The biotrophic pathogens seem to follow a somewhat different pathway. They follow the evolution of their hosts and in doing so seem to be more or less specialized most if not all the time. In the rusts, Leppik (1965) concluded that the available phytogenetic evidence, morphological data, and biological specialization, were indicative of the considerable age of this group of parasites. Their ancestors presumably already lived as facultative parasites on ferns and mosses in the Palaeozoic. This might resemble the initial phase as described in the model of coevolution, from which the rusts evolved following the evolution and differentiation of their host plants. If only an incidental evolution with their hosts occurred, as suggested in the model, it is difficult to explain how a relatively recent genus like *Puccinia* is found on host genera throughout the Angiospermae. *Erysiphe* is a similar example.

Observations of this type can be explained by 'jumping' as there are many indications that pathogens can jump to new and often distantly related hosts; this was called biogenic radiation by Leppik (1967). Such jumps, Savile (1971, 1979*a*) describes several, are difficult to explain with our present knowledge and do not fit very well into the model. Apparently, some pathogens have the capacity to break through the general defences of non-host plants occasionally, opening new possibilities for parasitism. Apart from these jumps to more or less distantly related host taxa, the biotrophic pathogens, being completely dependent on their hosts, must have their evolution finely tuned to those of their congenial hosts. It is most likely that coevolution plays a significant part in the process.

Coevolution does occur and may do so rapidly. In the rabbit-myxoma virus (Fenner 1965) the rabbit (*Oryctolagus cuniculus*) had become a serious pest in Australia but in 1950 the myxoma-virus, a natural pathogen of the Brazilian rabbit (*Sylvilagus brasiliensis*), was introduced into Australia. In the Brazilian rabbit it causes a relatively benign infection, but in Australia it was lethal, 99.5 per cent of the rabbits died. Eight years later the situation had dramatically changed. The rabbit population was considerably more resistant and the virus strains present appeared much less virulent than the one originally introduced resulting in a survival rate of 75 per cent, an enormous increase over the 0.5 per cent 8 years before. The rabbit-myxoma virus and its new host coevolved rapidly towards a coexistence (see Barrett, this volume).

Coevolution

The enormous diversification of present day pathogens seems to have developed through the combined effects of:
(a) adaptation to the host tissue parasitized (*specialization*);

(b) evolving in conjunction with the evolving host (*biological specialization*);
(c) occasional jumps to less or more distantly related hosts (*diversification*) into new taxa: biogenic radiation (Leppik 1967);
(d) evolution of pathogen and host in interdependence of each other (*coevolution*).

Coevolution almost unavoidably must occur once an intricate relationship has been established (Ehrlich and Raven 1964). In the coevolution model in this paper the first defence mechanisms are interpreted as of a general nature, the later ones more specific. Table 2.3 tries to depict this coevolution schematically and condensed into four steps. There is little doubt that this type of coevolution has occurred repeatedly, but it is very difficult to trace back the various small steps that have led to the present intricately and finely tuned coexistence of host and parasite. Concerning the steps 3 and 4 of the cereals and rusts, or more generally of the host plants and their biotrophic fungal pathogens, a large amount of research has been carried out showing the close coevolution that must have taken place.

Table 2.3. Coevolution between *Passiflora* species and *Heliconiini* butterflies and between cereals and their rust pathogens condensed into two steps (after Parlevliet 1983)

Step	*Passiflora* — *Heliconiini*	Cereals — *Puccinia* spp.
1	Toxic compounds effective to most insect species	Phyto-alexins against micro-organisms in general and produced when attacked or damaged
2	Detoxification by *Heliconiini*	Production of phyto-alexins suppressed by rusts
3	Small yellow protuberances imitating *Heliconiini* eggs on leaves or tendrils (*Heliconiini* lay only eggs on non-occupied plants)	Resistance genes effective to one rust species only
4	Some *Heliconiini* species inspect all yellow protuberances to identify truly unoccupied plants	Each rust species develops specific genes that neutralize the effects of the specific host resistance genes (gene-for-gene)

As soon as man started to use cultivars carrying single genes for resistance, variants of the pathogen neutralizing the effects of these resistance genes appeared (Barrus 1911). These variants were called physiological races (Stakman 1914) or, in short, races. Host species often appear to harbour large numbers of such genes conferring resistance to those highly specialized biotrophic pathogens like *E. graminis* f.sp. *hordei* (> 20 in barley), *P. graminis* f.sp. *tritici* (> 40 in wheat), and *Melampsora lini* (29 in flax). These resistance genes operate against one pathogen species only, they are not effective even against related pathogen species. The stem rust (*P. graminis*) resistance genes in wheat are only effective against races of stem rust, but not at all against races of leaf rust

(*P. recondita* f.sp. *tritici*), nor races of yellow rust (*P. striiformis*). In the pathogen population, genes are found that can fully neutralize the effect of such resistance genes. Those host resistance genes and pathogen virulence genes act in a gene-for-gene mode (Table 2.4). For each resistance gene in the host there is one specific virulence gene in the pathogen that can neutralize the effects of this resistance gene. In Table 2.4, the R-gene is neutralized by virulence gene a, the S-gene by virulence gene b. Flor (1956, 1971), in his classic study of the flax-rust (*Melampsora lini*) system, formulated this gene-for-gene relationship. The existence of gene-for-gene relationships have been proved in a range of host–pathogen systems (Day 1974). At present most scientists assume a gene-for-gene relationship to be at the basis of all major gene resistances to highly specialized pathogens.

Table 2.4. Reactions of 16 combinations between two host loci and two pathogen loci each with two alleles. The host genotypes have been taken as homozygous diploid, the pathogen genotypes haploid

Host	Pathogen genotype			
	AB	aB	Ab	ab
rr ss	+ *	+	+	+
RR ss	–	+	–	+
rr SS	–	–	+	+
RR SS	–	–	–	+

* + stands for a compatible reaction (susceptibility of the host, virulence of the pathogen); – for an incompatible reaction (resistance of the host, avirulence of the pathogen)

Studying these resistance genes a few interesting aspects occur.
(a) They are often numerous
(b) They are mostly dominant or incompletely dominant
(c) They tend to be clustered, either linked into groups of loci, very tightly linked in complex loci, or occurring as truly multiple allelic series.

Barley carries considerably more than 20 resistance genes against *E. graminis* f.sp. *hordei*. All the known and identified genes are located on one arm of chromosome 5 (5 loci) or on one arm of chromosome 4 (2 loci) (Jensen and Jørgensen 1975). On the Ml^{-a} locus at least 11 alleles have been identified, each resistance allele giving resistance to another spectrum of races. It is not certain yet whether this locus is a single locus with a series of multiple alleles or a complex locus, consisting of several tightly linked loci. A similar situation is reported from maize with regard to the maize rust (*P. sorghi*) where at least 25 resistance genes have been identified located at 6 loci. On the distal end of the short arm of chromosome 10, three of these loci (Rp5, Rp1, and Rp6) are found within 5 centimorgans of each other (Saxena and Hooker 1968). On the locus Rp1, 14 resistance alleles occur. Detailed studies revealed that this is not a single locus, but a complex one consisting of at least three very tightly linked loci with crossing over percentages in the order of 0.2-0.4 per cent. The Rp3 locus (chromosome 3) seems to carry a true multiple allelic series with six resistance alleles (Saxena and Hooker 1974). Such multiple allelic series and/or complex loci (difficult to distinguish) occur frequently and it is tempting to see a sequence of related resistance genes evolved in time in response to the evolving virulence of the pathogen.

It is now generally accepted (Ellingboe 1979; Parlevliet 1981; Daly 1984) that the gene-for-gene relationship as described in Table 2.4 results from the specific recognition of the products of the host resistance genes and the products of the avirulence genes of the pathogen superimposed upon basic compatibility. The gene products of R and A, and the gene products of S and B in Table 2.4 are assumed to fit each other as a lock and its corresponding key. When they match, a chain of reactions is started with incompatibility as the result. A change in either the product of the resistance gene or in the product of the avirulence gene may prevent the recognition and so the incompatibility. In other words, the basic compatibility is restored. In Table 2.4 the sequence might be interpreted as follows: a host cultivar is resistant because of gene *R* being fully effective to a pathogen consisting only of *AB* genotypes. A change in *A*, preventing the incompatibility reaction, gives again compatibility; this is pathogen genotype *aB*. If at some time *R* mutates so as to match this virulence gene *a*, derived from virulent *A*, again there is incompatibility. A new change in *a* again moves the pathogen towards virulence, and so on. Such mutations might easily lead to multiple allelic series and through duplications to loci adjacent to each other, each carrying one to several resistance alleles, the complex loci.

The result of the above assumed coevolutionary sequence is demonstrated with the flax-flax rust relationship. This is the most comprehensive study of the genetics of the interaction occurring between a plant and its pathogen and is especially due to the work of Flor (1956, 1971) who studied 26 resistance genes and the corresponding virulence genes. At present 29 resistance genes are known and for each of these host genes the pathogen possesses a specific gene conferring avirulence (Lawrence *et al.* 1981*a*). Host resistance is dominant to susceptibility and pathogen virulence nearly always recessive to avirulence. The 29 resistance genes occur on five loci, or rather on five tiny regions of chromosome, designated K, L, M, N, and P; these carry 1, 13, 7, 3, and 5

Table 2.5. Tightly linked groups of genes for resistance in the flax, *Linum usitatissimum*, and for avirulence in the flax rust, *Melampsora lini*

Host resistance genes	Pathogen avirulence genes
K	
L, L1, L2 . . . . L12	A-L3*, A-L4, A-L10
	A-L5, A-L6, A-L7
M, M1 . . . . M6	A-M1, A-M4
N, N1, N2	
P, P1 . . P4	A-P, A-P1, A-P2, A-P3

*A-L3 stands for avirulence to resistance gene L3.

resistance genes, respectively (Table 2.5). The fine structure has been studied by Shepherd (1963), Flor (1965b), Shepherd and Mayo (1972), and Mayo and Shepherd (1980). The 13 resistance genes on locus L appeared to belong to the same cistron; they are functional alleles. The M and N loci, however, are complex loci; the resistance genes in these groups are probably not functional alleles, but belong to separate, tightly-linked cistrons. For the M-region the linear order of four of the resistance genes for instance was found to be M, M4, M3, M1 (Mayo and Shepherd 1980). The P-region has not yet been studied in detail.

The corresponding avirulence genes in the pathogen have been studied in less detail, predominantly because of the greater difficulties in the genetic studies of avirulence. However, the results indicate that the avirulence genes exhibit less grouping than the resistance genes. In fact 12 avirulence genes appeared to occur in tightly linked groups as shown in Table 2.5 (Flor 1955; 1965a; Lawrence et al. 1981a). Recombination between genes within these four groups has not yet been reported. As only fairly small numbers of progeny were studied it is not possible to distinguish a tight linkage of loci from true multiple allelism in those groups. However, Shepherd developed a new selection procedure allowing the study of much larger numbers of progeny (Lawrence et al. 1981b). From this study it was concluded that the four avirulence genes A-P, A-P1, A-P2, and A-P3, probably belong to one and the same cistron, representing a true multiple allelic series. Of the other avirulence genes, A-L, A-L1, and A-L2 showed independent segregation among themselves and with the avirulence genes of the four groups. A-L8 appeared loosely linked with the A-L3, A-L4, A-L10 linkage group. A-N1 and A-N2 seem linked with each other, but to inherit independently of A-N.

The avirulence genes that were found to be linked together closely all correspond with resistance genes that are tightly linked or allelic suggesting a marked resistance-for-avirulence step by step coevolution at the gene, or rather DNA, level.

Implications for taxonomy

The intimate association resulting from coevolution, as described above, suggests interesting possibilities for using host range specialization for taxonomic purposes. However, there are four reasons why some caution is advisable.

(1) Not all groups of pathogens are suitable for taxonomic purposes. Those that can be classified as generalists, even though they may have highly specialized off-shoots like *A. alternata* and *F. oxysporum*, do not seem effective tools in taxonomy.

(2) The taxonomy of the pathogens is in many taxa poorly developed. The taxonomy of viruses is still in its infancy. Also, plant bacterial taxonomy is still poorly developed and even in the far more advanced taxonomy of fungi considerable problem areas still exist. In many cases the true taxonomic boundaries are rather obscure when examined in detail. For instance the powdery mildew *Erysiphe polygoni* is superficially seen as a generalist, attacking large numbers of species, but more detailed studies have revealed that *E. polygoni* is either a group of fully separated species or a complex species. This is reflected in the notations *E. polygoni sensu lato* and *E. polygoni sensu stricto*; in the latter sense it has been found only on Polygonaceae in Europe (Boerema and Verhoeven 1976, 1979, 1980; Yarwood 1978). If the taxonomy of a pathogen is not clearly understood, it will be difficult to use it effectively in host plant taxonomy.

(3) The jumping behaviour, as described by Leppik (1967) and Savile (1971, 1979a), may cause unexpected deviations from host–pathogen taxa associations, which has little to do with the taxonomy of the hosts and may confuse the taxonomist.

(4) The specialization of the pathogen may be based only on one or a few specific characteristics of the hosts. If these characteristics originated independently in different host taxa the specialized pathogen may accept these different host taxa as suitable hosts. For example, in Ehrlich and Raven's (1964) discussion of the coevolution of butterflies and their host plants, the butterfly sub-family Pierinae is reported to feed on the closely related Capparidaceae and Cruciferae, but also on the unrelated Salvadoraceae. These three host plant families all contain mustard-oil glucosides, secondary substances used as a general defence mechanism against parasitic insects. The Pierinae not only adapted to these specific secondary substances, they also started to use them as a means to identify their hosts.

The aspects discussed above do warn against a too optimistic viewpoint in considering pathogens or parasites in general as an effective tool in plant taxonomy. However, an extensive knowledge of both the taxonomy of the host and the pathogen certainly can be helpful in the classification of plants. This is precisely what experienced students in this field have concluded (Savile 1971, 1979a, 1979b; Hijwegen 1979). On the other hand, it should be realized that, in classifying the pathogens, the host reaction has often been an important consideration, possibly because of lack of sufficient other data. Thus, in classifying the hosts we should not overlook their associations with certain groups of specialized pathogens.

References

Anikster, Y. and Wahl, I. (1979). Coevolution of the rust fungi on Graminae and Liliaceae and their hosts. *Ann. Rev. Phytopath.* 17, 367–403.

Armstrong, G. M. and Armstrong, J. K. (1975). Reflections on the wilt Fusaria *Ann. Rev. Phytopath.* 13, 95–103.

Barrus, M. F. (1911). Variation of varieties of beans in their susceptibility to anthracnose. *Phytopath.* 1, 190–5.

Boerema, G. H. and Verhoeven, A. A. (1976). Check-list for scientific names of common parasitic fungi. Series 2a: Fungi on field crops: beet and potato; caraway, flax and oilseed poppy. *Neth. J. Plant Path.* 82, 193–214.

—— and —— (1979). Check-list for scientific names of common parasitic fungi. Series 2c: Fungi on field crops: pulse (legumes) and forage crops (herbage legumes). *Neth. J. Plant Path.* 85, 151–85.

—— and —— (1980). Check-list for scientific names of common parasitic fungi. Series 2d: Fungi on field crops: vegetables and cruciferous crops. *Neth. J. Plant Path.* 86, 199–228.

Comstock, J. C. and Scheffer, R. P. (1973). Role of host-selective toxin in colonization of corn leaves by *Helminthosporium carbonum*. *Phytopath.* 63, 24–9.

Daly, J. M. (1984). The role of recognition in plant disease. *Ann. Rev. Phytopath.* 22, 273–307.

Day, P. R. (1974). *Genetics of Host–parasite Interaction*. Freeman and Company, San Francisco.

Dobzhansky, T. (1970). *Genetics of the Evolutionary Process*. Columbia University Press, New York.

Ehrlich, P. R. and Raven, P. H. (1964). Butterflies and plants: a study in coevolution. *Evolution* 18, 586–608.

Ellingboe, A. H. (1979). Inheritance of specificity: the gene-for-gene hypothesis. In *Recognition and Specificity in Plant Host–parasite Interactions* (eds J. H. Daly and I. Uritani), pp. 3–17, Japanese Scientific Press, Tokyo.

Eshed, N. and Dinoor, A. (1980). Genetics of pathogenicity in *Puccinia coronata*: Pathogenic specialization at the host genus level. *Phytopath.* 70, 1042–6.

Fenner, F. (1965). Myxoma virus and *Oryctolagus cuniculus*: two colonizing species. In *Genetics of colonizing species* (eds H. G. Baker and G. L. Stebbins), pp. 485–99. Academic Press, New York.

Flor, H. H. (1955). Host–parasite interaction in flax rust—its genetics and other implications. *Phytopath.* 45, 680–5.

—— (1956). The complementary genic systems in flax and flax rust. *Adv. Genet.* 8, 29–54.

―― (1965a). Inheritance of smooth-spore-wall and pathogenicity in *Melampsora lini*. *Phytopath.* 55, 724-7.
―― (1965b). Tests for allelism of rust-resistance genes in flax. *Crop Sci.* 5, 415-8.
―― (1971). Current status of the gene-for-gene concept. *Ann. Rev. Phytopath.* 9, 275-96.
Gordon, W. L. (1965). Pathogenic strains of *Fusarium oxysporum. Can. J. Bot.* 43, 1309-18.
Hijwegen, T. (1979). Fungi as plant taxonomists. *Symb. bot. upsal.* 22(4), 146-65.
Jensen, J. and Jørgensen, J. (1975). The barley chromosoom 5 linkage map. II. Extension of the map with four loci. *Hereditas* 80, 17-26.
Kiraly, Z., Klement, Z., Solymosy, F., and Vörös, J. (1970). *Methods in Plant Pathology*. Akadémiai Kiadó, Budapest.
Kohmoto, K., Scheffer, R. P. and Whiteside, J. D. (1979). Host selective toxins from *Alternaria citri. Phytopathology* 69, 667-71.
Lawrence, G. J., Mayo, G. M. E. and Shepherd, K. W. (1981a). Interactions between genes controlling pathogenicity in the flax rust fungus. *Phytopath.* 71, 12-9.
―― Shepherd, K. W. and Mayo, G. M. E. (1981b). Fine structure of genes controlling pathogenicity in flax rust, *Melampsora lini. Heredity* 46, 297-313.
Leppik, E. E. (1965). Some viewpoints on the phylogeny of rust fungi. V. Evolution of biological specialization. *Mycologia* 57, 6-22.
―― (1967). Some viewpoints on the phylogeny of rust fungi. VI. Biogenic radiation. *Mycologia* 59, 568-79.
Leyns, F., de Cleene, M., Swings, J. G. and de Ley, J. (1984). The host range of the genus *Xanthomonas. Bot. Rev.* 50, 308-56.
Mayo, G. M. E. and Shepherd, K. W. (1980). Studies of genes controlling specific host-parasite interactions in flax and its rust. 1. Fine structure analysis of the M group in the host. *Heredity* 44, 211-27.
Mitchell, R. E. (1984). The relevance of non-host-specific toxins in the expression of virulence by pathogens. *Ann. Rev. Phytopath.* 22, 215-45.
Nishimura, S., Kohmoto, K. and Otani, H. (1979). The role of host-specific toxins in saprophytic pathogens. In *Recognition and Specificity in Plant Host-parasite Interactions* (eds J. M. Daly and I. Uritani), pp. 133-46. Japanese Scientific Press, Tokyo.
Parlevliet, J. E. (1979). The coevolution of host-parasite systems. *Symb. bot. upsal.* 22(4), 39-45.
―― (1981). Disease resistance in plants and its consequences for breeding. In *Plantbreeding* II (ed. K. J. Frey), pp. 309-64. Iowa State University Press, Ames.
―― (1983). Coevolutie en de implicaties ervan voor de plantenveredeling (Coevolution and its implications for plant breeding). In *Plantenveredeling en genenmanipulatie* (eds H. F. Linskens, H. M. van Emden, and R. J. P. Aalpol), pp. 62-75. PUDOC, Wageningen.
―― (1985). Race and pathotype concepts in parasitic fungi. *EPPO Bulletin* 15, 145-50.
Savile, D. B. O. (1971). Coevolution of the rust fungi and their hosts. *Q. Rev. Biol.* 46, 211-8.

—— (1979a). Fungi as aids to plant taxonomy: methodology and principles. *Symb. bot. upsal.* 22 (4), 135-45.
—— (1979b). Fungi as aids in higher plant classification. *Bot. Rev.* 45, 377-503.
Saxena, K. M. S. and Hooker, A. L. (1968). On the structure of a gene for disease resistance in maize. *Proc. natn Acad. Sci.* 61, 1300-5.
—— and —— (1974). A study on the structure of gene $Rp3$ for rust resistance in zea mays. *Can. J. Gen. Cyt.* 16, 857-60.
Scheffer, R. P. and Livingston, R. S. (1984). Host selective toxins and their role in plant diseases. *Science, N.Y.* 223, 17-21.
Shepherd, K. W. (1963). *Genetics of host-pathogen interactions.* PhD thesis, University of Adelaide.
—— and Mayo, G. M. E. (1972). Genes conferring specific plant disease resistance. *Science, N.Y.* 175, 375-80.
Stakman, E. C. (1914). A study in cereal rusts: physiological races. *Bull. Minn. Agric. Exp. Sta.* 138, 1-58.
Yarwood, C. E. (1978). History and taxonomy of powdery mildews. In *The Powdery Mildews* (ed. D. M. Spencer), pp. 1-37. Academic Press, London.
Yoder, O. C. (1980). Toxins in pathogenesis. *Ann. Rev. Phytopath.* 18, 103-129.

3. Aphid–plant associations

V. F. EASTOP

Department of Entomology, British Museum (Natural History), U.K.

Abstract

The host ranges of the major groups of aphids are summarized. More favoured and less favoured groups of plants are compared, as are the biologies of aphids on different groups of plants and in different geographical regions. Anomalous host plant records and problems in aphid phylogeny are considered.

Introduction

The extent to which the relationship between phytophagous insects and their plant hosts can be regarded as 'coevolution' is controversial. Futuyma (Futuyma and Statkin 1983) has reviewed the controversy.

The large number of host specific insects, the existence of 'secondary plant substances' and of more evident defence mechanisms such as sticky or hooked hairs, latex, moats formed by connate leaves, smooth shiny leaves and stems, etc., has led by a process of intuitive deduction to a concept of coevolution between plants and phytophagous insects. Gillett (1962) regarded 'pest pressure' as an important factor in the evolution of plants, and Ehrlich and Raven's (1965) title *Butterflies and Plants: a Study in Coevolution* is self-explanatory. Jermy (1976, 1984, p. 623), however, takes the view that while the biochemical and structural diversity of the angiosperms provide a profusion of niches for the evolutionary radiation (cladogenesis) of the insects, insects do not affect plant evolution or, at most, may cause anagenic changes in the plants. The direct unequivocal evidence for plant–insect coevolution is limited (Jermy 1976).

Jermy suggests that while plants may evolve defence mechanisms which

©The Systematics Association, 1986. This chapter is from *Coevolution and Systematics* (eds A. R. Stone and D. L. Hawksworth) published for the Systematics Association by the Clarendon Press, Oxford.

may be countered by insects, these mechanisms have not been important in the macroevolution of plants. The development of pollination mechanisms, on the other hand, may have had a considerable effect on the macroevolution of flowering plants, but not perhaps direct effects on the macroevolution of the insects and other pollinators. Any isolating mechanism of course may be of great importance indirectly. It is more difficult to appreciate the long-term effects of antagonistic than of collaborative processes. Armoured fighting vehicles and armour piercing shells are analogous (Eastop 1981a), successive developments in one having triggered successive developments in the other — the well known 'arms race' hypothesis of the coevolution of mutual aggression (see Holmes 1983).

The basic questions are whether any aspects of antagonistic relationships can be classified as coevolution, and if so, whether this coevolution had any major long-term effect on the evolution of either participant. We may observe that an insect survives on leaves bearing sticky hairs by walking slowly on small tarsi. We can assume that the density of sticky hairs increased with time, and that the insects learned to walk more carefully, and that the tarsi contracted with time. However, the plant may originally have been defending itself against a different aphid (or other insect) lineage, and the predecessors of the present aphid may have preadapted on a quite different sticky host. There has most probably been some reciprocal adaptation between interacting lineages, but it is difficult to prove, it is just more likely than not.

Aphids living on plants with evident defence mechanisms, sticky hairs, and smooth shiny stems and leaves, may be modified to deal with these conditions. The distribution of gall formation by aphids suggests a long relationship between some lineages. Groups of plants without or with very few aphids may be similar to those that support many, and historical reasons for the absence of aphids must be considered. However, analysis of these data are fraught with difficulties. Many other major influences are involved and some real effects are difficult to detect. A change in the chemistry or structure of a plant may result in a change of behaviour in the insect feeding on it. A restricted feeding site may be selected because it has most translocated nutrients (e.g. flower stalks) or may be the result of partitioning a habitat with other insects.

Aphids are obligate parasites of plants. About 4000 species of aphids are known and the very great majority (99 per cent) have an evidently restricted range of host plants, which in the majority of cases is expressed as host plant specificity at about the subgeneric level. About 10 per cent of aphids alternate between a usually woody primary host, which bears the sexual generations and their resultant overwintering eggs, and usually only distantly related herbaceous secondary hosts on which the parthenogenetic summer generations occur. Host-alternating aphids are highly specific to their primary hosts, and mostly also to their secondary hosts. There are about 25 aphid species with a wider host-plant range, mostly able to feed on many members of one plant family, and another ten aphid species which can feed on a wide range of plants in many families (Blackman and Eastop 1984, pp. 204–5). Another four aphid species have a wide taxonomic range of hosts on the young shoots of tropical and subtropical trees and shrubs. All these aphids with a wide host range are known as pests, and a more detailed account of their host range is given by Eastop (1981*b*).

The publication by Remaudière and Stroyan (1984, pp. 100–1) of a new subfamily classification of aphids makes a new analysis of the host association of the major groups of aphids opportune. Shaposhnikov (1985) considers the wider ramifications and implications of insect–plant relationships. He points out that the perfection of the perianal hair ring in root-feeding aphids led to a closer association between ant and aphid, which affected the body shape of the aphid and led to the ants utilizing fewer other food sources. Ant attendance may accelerate aphid feeding which affects the plants directly, and the absence of honeydew and activity of ants affects the chemistry and structure of the soil.

Many aspects of the relationships between plants and insects have been reviewed during the last few years. The works of Finch (1980) and Chapman *et al.* (1981) on chemical attraction; van Emden (1973) and Forey (1981) on relationships; Hedberg (1980) on the use of parasites in plant taxonomy; Jermy (1984) on the evolution of relationships; and Strong *et al.* (1984) on patterns and mechanisms, contain numerous references to sources of data. References to and details of the host relationships of aphids can be found in Blackman and Eastop (1984), and Eastop (1972, 1977, 1978, 1980). There are often doubts about the accuracy of the host plant data. When a few specimens of a known polyphagous species are wrongly described as a new species, it is inevitably listed as specific to the plant from which it was described. Conversely, many records may be disregarded as evidently based on misidentifications or abnormal occurrences. Some spurious records have probably not yet been detected. Doubts arise when we find the genus *Kaburagia* with eight species, five species described from Juglandaceae, one described from a single sample from *Ailanthus* (Simaroubaceae) only distantly related to Juglandaceae, but with similar leaves, and another from a sample from *Cassia* (Leguminosae) also with pinnate leaves, somewhat similar to those of some Juglandaceae. However, one species collected from *Quercus acutissima* by an experienced collector indicates that members of the genus do live on plants other than Juglandaceae.

Host ranges of the major groups of aphids

The host ranges of the major groups of aphids arranged according to the subfamily classification prepared by Remaudière and Stroyan (1984)

Table 3.1. Major groups of aphids and their hosts

		Host plants		Number of
		Primary	Secondary	Geographical distribution	species

	Primary	Secondary	Geographical distribution	Number of species
Aphididae Pemphiginae				
1. Eriosomatinae				
Pemphigini				
Pemphigina	Ulmaceae	roots of dicotyledons and Gramineae	Holarctic, Oriental	63
Prociphilina	Various dicotyledonous trees and shrubs	roots of Coniferae	Holarctic, Oriental	65
Pemphigina	Populus	roots of herbs	Holarctic, Oriental	81
Fordini	Anacardiaceae			
Fordina	Pistacia	roots, mostly Gramineae	mostly Middle Eastern	48
Melaphidina	Rhus	mosses	mostly Oriental	9
2. Mindarinae	Pinaceae	autoecious	Holarctic	4
3. Hormaphidinae				
Hormaphidini	Hamamelis	Betula	Holarctic	8
Nipponaphidini	Distylium	Fagaceae and Lauraceae	Oriental	82
Cerataphidini	Styrax	Gramineae and Palmae	Oriental	81
4. Tamaliinae	Arctostaphylos	autoecious	Neoarctic	4
5. Neophyllaphidini				
Neophyllaphis	Podocarpaceae and Araucariaceae	autoecious	Worldwide except Holarctic	11
Ceriferella	Epacridaceae	autoecious	Australia	1/2
6. Phloeomyzinae	Populus	autoecious	Holarctic	1/3
7. Lizeriinae	Dicotyledonous trees	autoecious	Southern Hemisphere	50
8. Greenideinae	Dicotyledonous trees	autoceious	Oriental and southern	127

Table 3.1. (*continued*)

	Host plants		Geographical distribution	Number of species
	Primary	Secondary		
9. Anoeciinae				
Aiceonini	Lauraceae	? autoecious		12
Anoecini	*Cornus*	roots of Gramineae		20
10. Thelaxinae	trees, mostly Betulaceae, Fagaceae and Juglandaceae	autoecious	Holarctic, Oriental	16
11. Phyllaphidinae	Mostly Fagaceae and Betulaceae	autoecious	Holarctic, Oriental	260
12. Saltusaphinae	Cyperaceae	autoecious	mostly Holarctic	70
13. Macropodaphidinae	*Artemisia, Dasiophora*	autoecious	Central Asia	5
14. Drepanosiphinae	*Acer*	autoecious	Holarctic, Oriental	38
15. Israelaphidinae	Gramineae	autoecious	Mediterranean	3
16. Chaitophorinae				
Chaitophorini		autoecious		
Chaitophorina	Saliceae	autoecious	Holarctic	85
Periphyllina	*Acer*	autoecious	Holarctic	34
Siphini	Gramineae	autoecious	mostly Western Palaearctic	22
17. Lachninae				
Cinarini	Coniferae	autoecious	Holarctic	268
Lachnini	Mostly trees	autoecious	Holarctic	47
Tramini	roots, mostly Compositae	autoecious	Palaearctic	32
18. Pterocommatinae	Saliceae	autoecious	Holarctic	44
19. Parachaitophorinae	*Spiraea*	autoecious	Nearctic	1

Continued overleaf

Table 3.1. (continued).

	Host plants		Geographical distribution	Number of species
	Primary	Secondary		
20. Aphidinae				
Aphidini				
Rhopalosiphina	Rosaceae	Gramineae	Holarctic	80
Aphidina	Dicotyledonous shrubs	mostly Rosidae and Asteridae	Holarctic	515
Macrosiphini	often Rosaceae	mostly higher dicotyledons and Gramineae	Holarctic	1634
Phylloxeridae	mostly Juglandaceae and Fagaceae	autoecious as far as known	Holarctic	69
Adelgidae				
Pineini	*Picea*	other Coniferae		
	Picea	*Pinus*	Holarctic	18
Adelgini	*Picea*	*Abies, Larix, Pseudotsuga*	Holarctic	29

are indicated in Table 3.1. Most groups have evident host plant associations. Of the 20 subfamilies of Aphididae recognized, only four contain species with host plant alternation. The implications of host alternation to the concept of coevolution are discussed on pp. 45–9. All members of the remaining 16 subfamilies spend the whole year on the same or closely related species of plants, and the great majority of these are trees and shrubs. Coniferae, Fagaceae, Betulaceae, Ulmaceae, Juglandaceae, Saliceae, Aceraceae, and Gramineae are the most favoured (Table 3.2). The older groups, Coniferae-Juglandaceae, predominate, but four major groups of aphids are associated with Saliceae and two each with Rosaceae, Aceraceae and Anacardiaceae. It is evident from Table 3.1 and from more detailed information (Eastop 1972) that some plant families host numerous species of aphids while others have none (Table 3.4) or few (Table 3.3) aphids specific to them. Some of these plant families with few aphids specific to them are not intrinsically unsuitable as aphid food, and are readily colonized by polyphagous aphids. Polyphagous aphids can be major pests of crops belonging to little favoured families, e.g. Malvaceae, Cucurbitaceae, and Solanaceae. Many of the more important virus diseases of potatoes are transmitted by aphids.

The higher dicotyledons do not have major groups of aphids associated with them, although there are many genera of Aphidinae associated with

Table 3.2 Most favoured aphid host–plant families

Host plants	Number of species	Distribution and growth form	Number of species of aphids
Coniferae	400	Temperate and subtropical trees	365
Hamamelidaceae	97	Subtropical trees and shrubs	22
Ulmaceae	175	Tropical and temperate trees and shrubs	54
Juglandaceae	50	Temperate and subtropical trees	55
Fagaceae	750	Cosmopolitan trees	211
Betulaceae	157	Temperate and mountain trees and shrubs	108
Saliceae	530	North temperate trees, shrubs	216
Grossulariaceae	240	Temperate shrubs	43
Rosaceae	2500	Trees, shrubs, perennial herbs	291
Elaeagnaceae	50	North hemisphere shrubs, steppes and coasts	7
Cornaceae	95	Temperate and montane trees and shrubs	14
Caprifoliaceae	420	North temperate mountains, many xerophytes, trees, shrubs, lianes	45

Table 3.3 Large plant families moderately favoured by aphids

	Number of species of plants	Distribution and growth form	Number of species of aphids
Lauraceae	2200	Tropical and subtropical trees and shrubs	24
Ranunculaceae	800	North temperate herbs	38
Moraceae	1400	Tropical and subtropical trees and shrubs	24
Caryophyllaceae	1800	Cosmopolitan herbs	28
Polygonaceae	800	North temperate herbs	57
Plumbaginaceae	530	Maritime and steppes herbs and shrubs	7
Cruciferae	3000	North temperate herbs	38
Ericaceae	1900	Cosmopolitan temperate shrubs	45
Saxifragaceae	640	North temperate herbs	9
Onagraceae	650	Temperate and tropical herbs	26
Combretaceae	550	Tropical and subtropical trees and shrubs	11
Rhamnaceae	900	Cosmopolitan trees and shrubs	20
Vitidaceae	700	Tropical and subtropical climbing shrubs	10
Burseraceae	550	Shrubs and trees	10
Anacardiaceae	600	Tropical and warm temperate trees and shrubs	55
Geraniaceae	800	Cosmopolitan herbs	9
Balsaminaceae	470	Holarctic and African herbs	11
Araliaceae	700	Tropical trees and shrubs	8
Umbelliferae	2900	North temperate herbs	107
Labiatae	3300	Cosmopolitan herbs and undershrubs	68
Plantaginaceae	260	Cosmopolitan herbs	9
Valerianaceae	380	Holarctic and African herbs	10
Compositae	19 000	Herbs, widespread	605
Gramineae	9000	Grasses, widespread	242
Total	53 830		1500

Compositae, others with Labiatae and with Caprifoliaceae, and other aphid genera live only on Scrophulariaceae, Rubiaceae, and Boraginaceae. Trees belonging to the higher dicotyledons (e.g. Theales) have few aphids specific to them. In the subfamily Hormaphidinae the tribe Cerataphidini utilize Styracaceae as primary hosts, but the other two tribes in the subfamily both migrate from Hamamelidaceae. Neither ferns nor mosses have primitive aphids living on them, their aphid fauna belongs to Rosaceae-feeding and Gramineae-feeding lineages. Some of the Coniferous host associations are probably old (e.g. *Prociphilina*, Mindarinae, *Neophyllaphis*, Adelgidae), but the Cinarini may be derived from Fagaceae-feeders and *Elatobion* from *Picea* has close relatives on Berberidaceae and together with *Sanbornia* and *Siphonatrophia* from Cupressaceae, evidently belong to a dicotyledon-feeding lineage.

The number of species of an insect group which are specific to particular families of plants is neither simply related to the number of species in the plant family, to the area covered by that family, nor to any other single factor. Different families of insects favour different families of plants. Eastop (1972, pp. 80-1) tabulated the host ranges then known for aphids, psyllids, aleyrodids, and diaspid scale insects. The six groups of plants hosting most species in each group of insects are shown in Table 3.5.

It will be seen that aphids favour Compositae and Coniferae, while psyllids, aleyrodids and diaspid scale insects favour Leguminosae and Myrtaceae, to some extent reflecting the vegetation of the areas where these groups are most numerous. Many aphids and psyllids are found on Compositae and Saliceae which are not much favoured by aleyrodids or diaspid scale insects. Aphids and aleyrodids are often found on Gramineae, but psyllids and diaspid scales rarely colonize grasses.

Aphids and diaspid scales are commoner on Northern Hemisphere plants and psyllids and aleyrodids are relatively commoner on Southern Hemisphere plants. Similarly, aphids are relatively much commoner on the plants characteristic of temperate regions, while psyllids are relatively commoner on families of predominantly tropical plants (Eastop, 1972, p. 175). The more recent detailed information concerning psyllids given by Hodkinson (1980), and White and Hodkinson (1985) confirms their southern affiliations.

A comparison of Table 3.2 with Table 3.4 shows that both the most favoured families and the families least favoured by aphids contain mostly trees and shrubs, but that the most favoured families occur mostly in temperate regions and on mountains in the tropics, while the least favoured families tend to be tropical, and if subtropical then from South Africa or New Zealand. The Proteaceae illustrate Gillet's (1962) concept of 'pest pressure'—the primitive members occurring in tropical rainforest, but the majority of species are xerophytic.

The Coniferae (400 species) and nine families of dicotyledonous trees and shrubs (with together 5164 species) host 1431 species of host specific aphids (Table 3.2). That is, families containing 5464 species of vascular plants (2.2 per cent of the world's total) host 32 per cent of the world's aphids. A further 23 families of dicotyledons (Table 3.3) and the Gramineae, together containing 44 820 species (18 per cent of the world's total) bear 1227 host specific aphids (28 per cent of the world's total). The Gramineae (9000 species) and Cyperaceae (4000) bear 242 and 74 species of aphids, respectively. That is, this 5 per cent of the world's vascular

Table 3.4. Large plant families without aphids specific to them

Piperaceae	1750 tropical shrubs or climbers
Cactaceae	2000 species, dry areas, mostly tropical America
Aizoaceae	1800 xerophytes, South African herbs, undershrubs
Amaranthaceae	878 tropical and temperate herbs and shrubs
Flacourtiaceae	1100 tropical and subtropical trees and shrubs
Begoniaceae	860 perennial herbs, especially America
Sapotaceae	800 tropical trees
Myrsiniaceae	1000 tropical and subtropical trees and shrubs, South Africa, New Zealand
Melastomataceae	3500 tropical and subtropical herbs, shrubs and trees
Proteaceae	1200 mostly xerophytes, but primitive species are rainforest trees
Meliaceae	1400 trees and shrubs in warm regions
Oxalidaceae	900 tropical and subtropical perennial herbs
Bignoniaceae	2500 trees, shrubs and lianes, mostly tropical

plants bears 7 per cent of the world's aphids. The remaining 65 per cent of the world's vascular plants bears 33 per cent of the world's aphids.

There are 150 families of dicotyledons together containing 7000 species, and 42 families of monocotyledons together containing 3134 species, without any aphids specific to them. Some of the families little colonized by aphids are colonized by members of the related group Psylloidea (Eastop 1972).

There is a striking difference between the Rosaceae with a relatively large aphid fauna and the much larger family Leguminosae with a much smaller aphid fauna. Rosaceae are often utilized as primary hosts by aphids migrating to herbs in the summer. Although Leguminosae are the secondary hosts of a few North American aphids (*Nearctaphis*) alternating from Rosaceae, aphids laying their eggs on Leguminosae seem never to have developed host alternation, or conversely, aphids with host alternation have never transferred to Leguminosae. The tendency for winged aphids leaving Rosaceae to fly to members of other families and for those leaving Leguminosae to colonize other legumes has considerable implications for virus transmission (Eastop 1983) and could thus have an indirect influence on the evolution of plant families.

Most host-alternating aphids show considerable specificity to both primary and secondary hosts. For instance, *Myzus lythri* alternates between *Prunus mahaleb* and *Lythrum salicaria*; *Hyperomyzus lactucae* (despite its name) alternates from *Ribes nigrum* to *Sonchus asper* and *S. oleraceus*; *H. pallidus* alternates between *R. grossularia* and *S. arvensis*; other *Hyperomyzus* and relatives alternate from *Ribes* spp. to other Liguliflorous Compositae and some to Scrophulariaceae. The alternate hosts of heteroecious aphids then known were cross-indexed by Eastop (1977, pp. 58–61). Generally,

Table 3.5. Most favoured hosts of aphids, psyllids, aleyrodids, and diaspid scales

Aphids		Psyllids		Aleyrodids		Diaspid scales	
Compositae	605	Myrtaceae	212	Leguminosae	58	Leguminosae	74
Coniferae	363	Leguminosae	125	Myrtaceae	43	Myrtaceae	46
Rosaceae	293	Compositae	84	Lauraceae	42	Coniferae	45
Gramineae	242	Saliceae	71	Moraceae	42	Fagaceae	37
Saliceae	216	Anacardiaceae	32	Gramineae	38	Euphorbiaceae	32
Fagaceae	211	Moraceae	29	Euphorbiaceae	25	Lauraceae	22

primary and secondary hosts are only very distantly related taxonomically. Hille Ris Lambers (1980, p. 115) discusses the various aphid species alternating from apple to different secondary hosts and concludes that this range of only distantly related secondary hosts is evidence that they were acquired relatively recently, rather than primitively retained.

Host ranges in different geographical areas

Most widespread groups of aphids have much the same host range throughout their geographical range. The Prociphilina on Oleaceae, Rosaceae and the roots of Conifers, Pemphigina on *Populus*, Mindarinae on Coniferae, *Neophyllaphis* on Podocarpaceae and Araucariaceae, Phyllaphinae on Fagaceae, Betulaceae, Ulmaceae, etc., Saltusaphidinae on Cyperaceae, Drepanosiphinae on *Acer*, *Chaitophorina* on Saliceae, *Periphyllina* on *Acer*, Pterocommatinae on Saliceae, Cinarini on Pinaceae and Cupressaceae, Rhopalosiphina on Rosaceae and Gramineae and Cyperaceae. However, in a number of cases very similar aphids from different continents feed on quite different plants. The African Lizeriinae live on Combretaceae and Burseraceae, but in South America they occur on a number of other families (Table 3.6). In the Thelaxinae (Table 3.7) the North Temperate species of *Thelaxes* and *Glyphina* live on Fagaceae and Betulaceae, respectively, but the oriental *Kurisakia* is mostly on Juglandaceae with a few species described from other plants, with similar looking leaves. Within genera there may be species groups in a particular continent with a characteristic host range. This apparently results from the ability to colonize a new habitat resulting in speciation- -e.g. the American *Sitobion*-like aphids on ferns or the African *Sitobion* on trees. In the Palaearctic region a subgenus of *Aphis*, *Pergandeida*, has a rather uniform appearance and is largely confined to Leguminosae. In South America the group is more variable structurally and also occurs on a wider variety of plants in different families.

While the availability of a different variety of plant families sometimes

Table 3.6. Paoliellinae and host plants

		Lauraceae	Nothofagus (Fagaceae)	Gunneraceae	Nyctaginaceae	Leguminosae	Myrtaceae	Combretaceae	Burseraceae	Rubiaceae	Total
Taiwanaphis	S. E. Asia Australia						1			1	2
Sensoriaphis	New Zealand New Guinea		4				1				5
Neuquenaphis	S. America		9	1							10
Lizerius	S. America	3			1			1			5
Paoliella	Africa							10	7		17
Antalus	Africa					2					2
Total		3	13	1	1	2	2	11	7	1	41

Table 3.7. Thelaxinae and host plants

	Juglandaceae	Fagaceae	Betulaceae	Leguminosae	Simaroulaceae	Total	Europe	Far East	North America
Glyphina			6			6	3	1	2
Thelaxes		4				4	2		2
Kurisakia	5	1		1	1	8		8	
Total	5	5	6	1	1	18	5	9	4

results in a wider host range, this is not common. It may of course be concealed in part by the host plant association influencing generic classification.

Most plant relationships are evidently influenced by geography and climate. Aphididae have a basic biology alternating viviparous parthenogenetic females in the spring and summer with a male and female in the autumn producing an overwintering egg for which a cold diapause-breaking spell may be needed.

Eastop (1972, p. 175) showed that the 30 000 species of plants belonging to exclusively tropical families hosted only 39 aphids (1.3 aphids per 1000 plants); the 55 000 members of tropical and subtropical families had 308 aphids (5.6 per 1000); the 94 600 plant species belonging to cosmopolitan families hosted 2227 aphids (24 per 1000) and the 14 300 members of temperate and mostly temperate families hosted 804 aphids (56 per 1000). The corresponding figures for psyllids are 1, 2, 6, and 7 per 1000 species of plants, and this underestimated the tropical fauna which is now better known.

Biologies of aphids on different groups of plants

About 10 per cent of all aphids, belonging to four subfamilies, alternate seasonally between members of two different families, but the remaining 90 per cent of species, including all the members of 16 subfamilies, remain on the same, or closely related hosts, throughout the year. This presents nutritional problems on some plants. In cold and temperate climates most aphids overwinter in the egg stage.

Overwintering on deciduous plants in temperate regions is achieved by eggs that are the progeny of sexuales induced by the lower temperatures and lengthening autumn nights. Aphids living on the leaves of deciduous trees during the summer may also face nutritional problems and many tree-leaf feeding aphids seem to be in reproductive diapause during mid-summer. Others, particularly *Acer*-feeding and *Quercus*-feeding species, may have special aestivating larvae in the summer which only complete their development in the autumn. Some conifer-feeding aphids found on the twigs in spring and autumn occur on roots in the summer. As some also occur under branches where these touch the ground, these aphids may be conserving moisture rather than seeking food.

A few aphids produce the overwintering egg very early in the season, e.g. *Dysaphis devecta* on apple; *Mindarus abietinus* on *Abies*; *Neophyllaphis totarae* on *Podocarpus totara*; *Sensoriaphis nothofagi* on *Nothofagus truncata* and *N. solandri*; and *Aphis farinosa* on *Salix*.

Even though the early production of sexuales may be triggered by the nutritative status of the plant as in *Dysaphis devecta* on apple (Forrest 1970) and most other apple-feeding aphids alternate to other plants (Hille Ris Lambers 1980, p. 115), a few species manage to survive on apple during the summer, either by feeding on the bark (*Eriosoma lanigerum*) or on the regenerated young growth (*Aphis pomi*). It seems that mature apple leaves are not liked by aphids. As the nutritative status of plants changes seasonally, overcoming these differences may be more difficult than changing to a different host species in a more suitable physiological condition.

Host specificity is generally lower in the monocotyledons than the dicotyledons. Monocotyledonous families and genera do not usually have such distinct aphid faunas as those of dicotyledons. For instance, the banana aphid, *Pentalonia nigronervosa*, lives on Musaceae, Araceae and members of a few other families and *Dysaphis tulipae* lives on Liliaceae, Iridaceae, and sometimes other families. Some grass-feeding aphids have a wide host range but a few species are host-specific (e.g. *Hyalopteroides pallida* on *Dactylis glomerata*). Bamboos have a distinctive aphid fauna, but seemingly not very specific within the Bambuseae. In contrast to the Gramineae, many Cyperaceae-feeding aphids are specific to particular *Carex* species. It is difficult to compare the effects on their hosts of polyphagous grass-feeders with specific sedge-feeders. Aphid-transmitted viruses of Gramineae such as barley yellow dwarf have a wide host range in both Gramineae and Cyperaceae, but almost only the virus diseases of economic plants are studied.

Anomalous host–plant relationships

Amphorophora is a well defined genus of about 27 species, mostly neoarctic, but with a few from the eastern and western palaearctic. Half of the species live on *Rubus* and a few on other Rosaceae such as *Geum* and *Filipendula*, but there are also species groups on ferns and on Geraniaceae. *A. stachyophila* from *Stachys rigida* is barely separable morphometrically from *A. rubitoxica* from *Rubus ursinus* and *R. vitifolia*, although the karyotypes of the two aphids are quite distinct. Most Tramini live on the roots of Compositae, but *Protrama ranunculi* from *Ranunculus* is very similar to *P. flavescens* from *Artemisia*. Most *Dysaphis* live on Umbelliferae including *D. foeniculi* and *D. crithmi*, which are similar to *D. ranunculi* from *Ranunculus*.

Nine of the ten species of the South American genus *Neuquenaphis* live on *Nothofagus* (Fagaceae), but *N. valdiviana* lives on *Gunnera* (Gunneraceae). *Neuquenaphis* are distantly related to *Sensoriaphis* known from Australia, New Guinea and New Zealand with four or five species on *Nothofagus* and one from *Melaleuca* (Myrtaceae). However, *Sensoriaphis* may not be more than subgenerically distinct from the oriental genus *Taiwanaphis*, with species on Myrtaceae and Rubiaceae.

Tables 3.6–3.8 show the host ranges of the genera of Paoliellinae, Thelaxinae, and Lachninae, tribes Lachnini and the related tribe Tramini in more detail. The general picture is similar in each case. Within each subgroup there are evident host preferences, but their arrangement does not suggest that the aphids evolved with plants as sedentary parasites, but that capture of new hosts, somewhat related to the old hosts, has often occurred. This picture is reinforced when the species in individual genera are concerned. For instance, in both the Saliceae-feeding *Chaitophorus* and *Pterocomma*, there are pairs of very similar species, one feeding only on a few species of *Salix* and the other on some species

Table 3.8. Host plants of Lachnini and Tramini

	Living on two or more plant families	Conifer roots	Altingiaceae	Ulmaceae	Fagaceae	Betulaceae	Tamaricaceae	Salicaceae	Rosaceae	Elaeagnaceae	Euphorbiaceae	Aceraceae	Compositae roots	Other families	Unknown	Total
Stomaphis	1	2	1	3	3	3		3			1	3		2[a]	2	24
Longistigma	2		1													3
Nippolachnus									3							3
Tuberolachnus								1	1							2
Pyrolachnus									3							3
Pterochloroides									1							1
Maculolachnus									2							2
Lachnus					9					1						10
Sinolachnus										1						1
Eotrama														1		1
Tramini							3				2		21	6[b]	1	33
Total	3	2	2	3	12	3	3	4	10	2	3	3	22	8	3	83

[a] One each from Juglandaceae and Hippocastanaceae.
[b] One each from Ranunculaceae, Polygonaceae, Cistaceae, Leguminosae, Umbelliferae, and possibly grass roots.

of *Populus*. Other Saliceae-feeding genera are restricted either to *Populus* (e.g. *Pemphigus*) or *Salix* (e.g. *Cavariella*).

Members of the genus *Pemphigus* form galls of *Populus* in spring and mostly migrate to the roots of secondary hosts during the summer. Many of the species are very similar to each other and the group is difficult taxonomically, and yet the secondary hosts include only distantly related families in different major groups of dicotyledons. It is difficult to see how a group of closely related species can alternate from a small group of related *Populus* species to a wide range of secondary hosts unless this wide range were acquired by capture. The common ancestor of *Populus*, *Chenopodium*, *Daucus*, and *Lactuca* lived too long ago to have been colonized by the aphid genus *Pemphigus*. There are two groups of genera within the Pemphigini, one containing *Pemphigus* and *Thecabius* alternates between galls on *Populus* and mostly herbs, while *Prociphilus* and allies alternate from a variety of trees and shrubs to the roots of Coniferae. It seems likely that the original Pemphigini alternated from a now extinct plant related to Ulmaceae and Hamamelidaceae to conifer roots, and that the *Prociphilus* group retained secondary hosts in the Coniferae while the *Pemphigus* group both acquired *Populus* as a primary host and new dicotyledons as secondary hosts.

Most aphid genera show an evident taxonomic pattern in their host plant relationships, but in a few cases the factors are evidently ecological or physiological. Most members of the subtribe Rhopalosiphina alternate between Rosaceae, and Gramineae or Cyperaceae, or are autoecious on one of the latter. *Rhopalosiphum nymphaeae*, however, lives on a variety of aquatic plants with large intercellular air spaces including ferns, monocotyledons and dicotyledons. *Rhopalosiphoninus latysiphon* and *R. staphyleae* are found on the aetiolated parts of a great variety of plants in cryptic habitats such as caves, or under stones or logs. Other *Rhopalosiphoninus* species such as *R. cathae* and *R. ribesinus* are, respectively, specific to *Caltha* and *Ribes* species, but are found in damp and shady places.

Host plant acceptability can be strongly modified by environmental conditions, by temperature or the presence of other aphids, which may encourage or deter other species. Plants growing rapidly in a cold glasshouse in the spring may be colonized by many species of aphids not usually found on them and similarly, the young growth on clipped hedges in early summer may be colonized by aphids not found under natural conditions. Kaltenbach (1843, pp. xxi–xxii) observed long ago that aphids do best on plants growing in rich soil and especially on plants which have been pruned hard. Eastop (1972, p. 164) gave a number of examples of aphids flourishing on plants on which they rarely occur. Hille Ris Lambers (1980, pp. 118–9) reported the colonization of *Matricaria* (Compositae) over a considerable area in the Netherlands, by *Cavariella aegopodii*, an aphid normally alternating between *Salix* and Umbelliferae.

Some plants seem particularly susceptible to a large range of aphids which usually live on other hosts. Many dicotyledon-feeding aphids will live on *Capsella* and/or *Rumex*, and many Gramineae-feeding aphids will live on *Poa annua*. Some aquatic plants, particularly *Polygonum hydropiper*, can be used to rear a variety of both dicotyledon-feeding and monocotyledon-feeding aphids. The importance of physiological conditions to host-acceptance by aphids is acknowledged by gardeners when they pick out the tops of broad beans to reduce infestation by black bean aphid.

Aphids living on particular plants tend to be modified in similar ways. Aphids living on Gramineae tend to have short, heart-shaped ultimate rostral segments. Aphids living on shiny leaves tend to have long second tarsal segments. Aphids on densely pubescent plants tend to have a long, slender ultimate rostral segment, and aphids on plants protected by sticky hairs usually have short tarsi. The processus terminalis and cauda of a number of moss-feeding genera of aphids have a characteristic structure.

Many aphids seem to have structural modifications associated with characteristics of their host plants. The ability to live on one plant species protected by a particular set of defences obviously facilitates colonization of other plants protected in the same way.

Gibson (1971) regarded the sticky hairs on several South American *Solanum* species as a defence mechanism against virus transmitting aphids, and sought to transfer these hairs to commercial potato varieties for protection from viruses. There are two problems with this attractive idea, firstly the species of *Solanum* concerned belonged to different groups, and secondly no South American *Solanum*-feeding aphids are known. These anomalies can be resolved by assuming that the ancestor of *Solanum*, *Nicotiana*, and *Petunia* was protected by sticky hairs and that these defences were lost by many of the species which evolved in the Andes where protection against aphids was no longer a great advantage.

Many plants bear branched or stellate hairs, and some hairy varieties of crops suffer less from insects and mites. Alatae of the aphid genus *Cervaphis* are unexceptional aphids, but the apterae of *Cervaphis* bear numerous long-branched spine-bearing projections. The leaves of their hosts are densely covered with stellate or branched hairs of about the same size. The branched projections of the aphids may act as tactile camouflage against predators such as dipterous larvae hunting by sense touch, which would be seeking spherical prey.

Most aphids are cryptic, but *Aphis nerii* is bright yellow and forms conspicuous colonies on the twigs of its poisonous host plants belonging to the Apocynaceae and Asclepiadaceae. The aphid sequesters the plant poisons (Brown *et al.* 1969; Rothschild *et al.* 1970) and is presumably protected by them. The ants attending the aphids may protect the plants from some other pests, but not evidently more than those attending inconspicuous aphids.

Plants have various structural defences against their enemies and also many more subtle mechanisms, such as seasonal variation in nutritative status. Gall formation usually involves a reaction of the plant to specific stimuli from insects which may include both active ingredients in the saliva and a pattern of feeding probes. Galls are often characteristic for a particular insect species and are usually regarded as evidence of an old host association. Gall-forming insects are often evidently modified for living in galls, and gall formation represents a sort of coevolution.

Phylogeny of Aphidoidea

Reconstructing the primitive aphids presents a variety of problems, but both Shaposhnikov (1985, pp. 53–9) and Heie (1985, pp. 104–5) have made the attempt. Within the Aphidoidea the Adelgidae(inae) and Phylloxeridae(inae) are so reduced as to be of limited help in reconstructing the primitive aphid and the probable sister group of the Aphidoidea, the Coccoidea are also greatly reduced, not having any winged females, and the males, even when winged, being evidently specialized and reduced. Shaposhnikov (1985) commented that 'path of evolution of aphids involve not so much acquisition of new structures, as transformation and not infrequently loss of the old ones'.

Heie (1985) summarized our knowledge of fossil aphids. *Triassaphis* from the Triassic may be very different from present-day aphids, but the Jurassic yields species similar to recent aphids.

The Cretaceous fossil *Oviparosiphum* is similar in many respects to *Neophyllaphis*, differing apparently in the structure of the secondary rhinaria and in the absence of siphunculi. The supposed ovipositor could be only a folded, enlarged sclerotized genital plate, such as is found in oviparae of *Neophyllaphis*, in which the siphunculi may be inconspicuous. Heie (1985, pp. 105) concludes that the Aphidoidea had been split into several groups before the Cretaceous period, pointing out that while present-day phylloxerids are easily distinguished from other aphid groups, they are characterized partly by plesiomorphous characters and partly by reduced or lost structures, which have been lost by other groups too, so it may be difficult or even impossible to recognize past members of the group.

Conclusions

Probably coevolution in the sense of the progressive development of plant defence mechanisms progressively countered by aphids, is common. These adaptations then enable aphids to live on quite different groups of plants protected by similar defence mechanisms. Several defence mechanisms seem to have arisen on many different occasions and there is little unequivocal evidence that the aphid/plant relationship has materially affected the macroevolution of either plants or aphids. However, evolutionary selection can only operate around the species level, and as plastic antagonistic defence mechanisms leave few traces, it is difficult to assess its importance in events that occurred from 20 to 200 million years ago.

References

Blackman, R. L. and Eastop, V. F. (1984). *Aphids on the World's Crops*. Wiley, Chichester.

Brown, K. S., Cameron, D. W., and Weiss, U. (1969). Chemical constituent of the bright orange aphid *Aphis nerii* Fonscolombe. I. Neriaphin and 6-hydroxymasizin 8-0-B-D-Glucoside. *Tetrahedon Lett.* 6, 471–6.

Chapman, R. F., Bernays, E. A. and Simpson, S. J. (1981). Attraction and repulsion of the aphid, *Cavariella aegopodii*, by plant ordors. *J. chem. Ecol.* 7, 881–8.

Eastop, V. F. (1972). Deductions from the present-day host plants of aphids and related insects. *R. ent. Soc. Symposium* 6, 157–78.

—— (1973). Biotypes of aphids. *Bull. Ent. Soc. N.Z.* 2, 40–51.

—— (1977). Worldwide importance of aphids as virus vectors (eds K. F. Harris and K. Maramorosch), pp. 3–62. Academic Press, London.

—— (1978). Diversity of the Sternorrhyncha within major climatic zones. *Symp. R. ent. Soc. Lond.* 9, 71–88.

—— (1980). Sternorrhyncha as angiosperm taxonomists. *Symb. bot. upsal.* 22(4), 120–38.

—— (1981a). Coevolution of plants and insects. In *The Evolving Biosphere* (ed. P. L. Forey), pp. 179–90. British Museum (Natural History) and Cambridge University Press, Cambridge.

—— (1981b). The wild hosts of aphid pests. In *Pests, Pathogens and Vegetation* (ed. J. M. Thresh), pp. 285–98. Pitman, London.

—— (1983). The biology of the principal aphid virus vectors. In *Plant Virus Epidemiology* (eds R. L. Plumb and J. M. Thresh), pp. 117–32. Blackwell Scientific Publications, Oxford.

Ehrlich, P. R. and Raven, P. H. (1965). Butterflies and plants: a study in coevolution. *Evolution* 18, 586–608.

van Emden, H. F. (1973). Aphid host plant relationships. *Bull. ent. Soc. N.Z.* 2, 54–64.

Finch, S. (1980). Chemical attraction of plant-feeding insects to plants. *Appl. Biol.* 5, 67–143.

Forrest, J. M. S. (1970). The effect of material and larval experience on morph determination in *Dysaphis devecta*. *J. Insect Physiol.* 16, 2281–92.

Futuyma, D. J. and Slatkin, M. (1983). *Coevolution*. Sinauer Associates, Sunderland, Massachusetts.

Gibson, R. W. (1971). Glandular hairs providing resistance to aphids in certain wild potato species. *Ann. appl. Biol.* 68, 113–9.

Gillett, J. B. (1962). Pest pressure, an underestimated factor in evolution. In *Taxonomy and Geography* (ed. D. Nichols), pp. 37–46. Systematics Association, London.

Hedberg, I. (ed.) (1980). Parasites as plant taxonomists. *Symb. bot. upsal.* 22(4), 1–221.

Heie, O. E. (1985). Fossil aphids. In *Evolution and Biosystematics of Aphids*, (ed. H. Szelegiewicz), pp. 101–34. Polish Academy of Sciences, Wroclaw.

Hille Ris Lambers, D. (1980). Aphids as botanists? *Symb. bot. upsal.* 22(4), 114–9.

Hodkinson, I. D. (1980). Present-day distribution patterns of the holarctic Psylloidea (Homoptera-Insecta) with particular reference to the origin of the Nearctic fauna. *J. Biogeog.* 7, 127–46.

Holmes, J. C. (1983). Evolutionary relationships between parasitic helminths and their hosts. In *Coevolution* (eds D. J. Futuyma and M. Slatkin), pp. 161-85. Sinauer Associates, Sunderland, Massachusetts.

Jermy, T. (1976). Insect-hostplant relationship—coevolution or sequential evolution? *Symp. Biol. Hung.* 16, 109-13.

—— (1984). Evolution of insect/hostplant relationships. *Am. Naturalist* 124, 609-30.

Kaltenbach, J. H. (1843). *Monographie der familien der Pflanzenläuse (Phytophthires).* P. Fagot, Aachen.

Remaudière, G. and Stroyan, H. L. G. (1984). Un Tamalia nouveau de Californe (USA) discussion sur les Tamaliinae subfam. nov. (Hom. Aphididae). *Ann. Soc. ent. Fr. (N.S.)* 20, 93-107.

Rothschild, M., Euw, J., and von, Reichstein, T. (1970). Cardiac Glycosides in the oleander aphid, *Aphis nerii. J. insect. Physiol.* 16, 1141-5.

Shaposhnikov, G. Ch. (1985). The main feature of the evolution of aphids. In *Evolution and biosystematics of aphids* (ed. H. Szelegiewicz), pp. 19-99. Polish Academy of Sciences, Wroclaw.

Strong, D. R., Lawton, J. H. and Southwood, T. R. E. (1984). *Insects on Plants. Community Patterns and Mechanisms.* Blackwell Scientific Publications, Oxford.

White, I. M. and Hodkinson, I. D. (1985). Nymphal taxonomy and systematics of the Psylloidea (Homoptera). *Bull. Br. Mus. nat. Hist.* (Ent.) 50, 153-301.

4. *Nothofagus* and its parasites: a cladistic approach to coevolution

C. J. HUMPHRIES
Department of Botany, British Museum (Natural History), U.K.

J. M. COX
Department of Entomology, British Museum (Natural History), U.K.

E. S. NIELSEN
Division of Entomology, CSIRO, Australia

Abstract

Cladistics is the discipline that hypothesizes genealogical relationships among taxa. In this paper we use parsimony methods to find the shortest possible sequence (principle of parsimony) of grouping character states (synapomorphies), to compare the phylogenies of the contemporary host species of *Nothofagus* (Fagales) and the species of three groups of contemporary parasites, *Heterobathmia* (Lepidoptera), Eriococcidae (Hemiptera), and *Cyttaria* (Cyttariales) which live on them. We consider two general questions. Firstly, do hosts and parasites share a common history of 'association by descent' (coevolution) or have parasites become parasitic through colonization. Secondly, we ask whether coevolution is a by-product of biogeography.

Introduction: parasitism by 'association by descent' and 'colonization'

Mitter and Brooks (1983) pointed out that the idea of coevolution by 'association by descent' is an old one (e.g. Kellog 1896). Examples from a variety of parasitological studies between hosts and parasites abound, especially internal parasites of fishes and tetrapods, and fungi on plants and animals. Among plants the southern beeches of *Nothofagus* are parasitized by a wide variety of different groups including insects, (e.g. moths, psyllids, coccids, and aphids), fungi, and some mistletoes. Schlinger (1974) gives a remarkable narrative with an example of an intricate phytophagous insect–primary parasitoid insect–hyperparasitoid insect interaction for *Nothofagus*, involving the phytophagous aphids *Neuquenaphis* and *Sensoriaphis*, their primary parasites *Pseudephedrus* and *Paraphedrus*, respectively, and their secondary parasites *Alloxysta*.

The concept that the systematic and biogeographic patterns of hosts might be more positively elucidated by looking at their specific parasites was articulated clearly by Metcalf (1920, 1929) who studied leptodactylid frogs and their opalinid parasites. Science proceeds with formal characterizations of problems and Eichler (1941) began to formalize Metcalf's ideas with 'Fahrenholz's rule' (see Mitter and Brooks 1983) which states that: 'in the case of permanent parasitism the relationship of the host can usually be inferred from the systematics of the parasites'. Hennig (1966) pointed out, however, that most studies utilizing this concept were based not on genealogies, but 'affinities'. Despite the detailed studies of Schlinger (1974) on the *Nothofagus* aphids and their parasites, where he acknowledges the importance of phylogenetic taxonomy, and those of Korf (1983) on the phylogeny of the parasitic *Cyttaria* fungi, both authors derive their coevolutionary theories from the *Nothofagus* phylogeny rather than using independent phylogenies and comparisons of the parasites and hosts.

We intend to extend beyond Fahrenholz's rule by using independent cladistic studies of three parasite groups, the moths of *Heterobathmia*, the scale insects of Eriococcidae, and the fungi of *Cyttaria* on their *Nothofagus* hosts. We try to see to what extent equivalent components occur in the hosts and their parasites rather than having one dependent on the other.

1. Cladistics: the method and an algorithm

Phylogenetic patterns are best represented by a branching diagram, a cladogram (Fig. 4.1). The tips or terminals (e.g. A, B, C, D, or 1, 2, 3, 4) represent the individual taxa from which the character data have been sampled. The nodes or branch points are determined by character state distributions (e.g. Fig. 1a–c). The character state polarities within each group are usually determined by comparison to an outgroup, a related group which may or may not be the sister group to the study group. One purpose of cladistics is to recognize monophyletic groups which for us are determined by proposing cladograms that have the (a)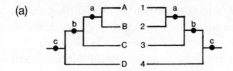

(b)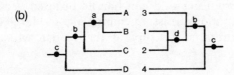

Fig. 4.1. Hypothetical phylogenies of a host group (A–D) and its parasites (1–4). (a) Exact correspondence between the host and parasite cladograms each showing the same components a, b and c. (b) Partial correspondence between the host and parasite cladograms each sharing two common components, b and c, and two unique components, a and d.

simplest (most parsimonious) distributions of character state changes over the taxa, i.e. the fewest number of proposed character state changes. For a fully dichotomized cladogram with n taxa there are $n-1$ nodes. The smallest number of changes for a specified characteristic is one less than the number of character states, i.e. a single origin for each state on the tree (Mitter and Brooks 1983). When a particular state is found in all and only the members of a putative monophyletic group, then we can propose that the character state evolved once and only once. In the general sense Hennig (1966) said that any group of taxa composed of all the descendents of a single ancestor, and only those descendents, is monophyletic, a definition which can be justified empirically by saying that monophyletic groups are due to common ancestry and homologous characters distinguish monophyletic groups (see Patterson 1982). For a more detailed discussion of cladistics with reference to the application of parsimony and an appraisal of homology see Humphries and Funk (1984).

In estimating a phylogeny for a particular group it is necessary to select one cladogram from a multitude of others. Most taxonomists are not aware of how many there are for a given number of taxa (Felsenstein 1978). For bifurcating rooted cladograms there are 3 for 3 taxa, 105 for 5 and 34459425 for 10! The computational problem is now simplified by several computer programs (see Luckow and Pimentel 1985). Those incorporating the Wagner algorithm (Farris 1970) or the branch and bound algorithm (Hendy and Penny 1982) are perhaps the most effective since they are both consistent with the parsimony criterion for minimizing the number of character state changes during analysis. Both algorithms find the cladogram of minimum length, the most parsimonious one from a given set of information (see Kluge and Farris 1969; Farris 1970). We used PAUP (Phylogenetic Analysis Using Parsimony), an interactive Fortran 77 program for inferring phylogenies under maximum parsimony (Swofford 1984). PAUP allows a choice of optimization procedures (accelerated transformations, delayed transformations, and Farris optimization), it holds equally parsimonious cladograms in memory, has branch-swapping routines and can handle ordered versus unordered character transformations, missing characters, and large data matrices. The computer printouts are kept at the British Museum (Natural History).

2. Components, coevolution and colonization

In comparing two independent phylogenetic hypotheses such as that of a parasite and that of its host we require a method based on the original empirical observations; the character data. Component analysis is one way this may be done (Nelson and Platnick 1981). In Fig. 4.1 the letters A–D refer to the host taxa in a host phylogeny and the numbers at the terminals refer to taxa of two different parasite phylogenies. The characters upon which each of the three phylogenies have been based can be substituted with letters a, b, c . . . etc. to define similar components. In Fig 4.1a complete correspondence between host and parasite is represented by the fact that components a–c, effectively hypothetical ancestors in this case, are common to both groups. The most defensible hypothesis, assuming the cladograms are equivalent to the actual host and parasite phylogenies, is that the ancestors of the groups A, B, C, D, and 1, 2, 3, 4 speciated at node 'C' to give two taxa for each group the ancestor of A, B, C, and D in the host, and 1, 2, 3, and 4 in the parasite. A second speciation event at node 'b' gave the ancestor of A, B, and C and 1, 2, and 3. The most parsimonious hypothesis explaining the exact correspondence between host and parasite is that speciation events occurred simultaneously (Mitter and Brooks 1983).

In Fig. 4.1b, the speciation events in common are represented by components 'b' and 'c', suggesting a partially shared history. However, non-correspondence between components 'a' and 'd' can only be explained by *ad hoc* hypotheses. If the host phylogeny is assumed to be 'correct', then the alternative component 'd' in the parasite phylogeny requires explanation. One possibility is that parasitism by colonization occurred; either species B was invaded by parasite 2 from C which then speciated or that species C was invaded by parasite 1 from B which subsequently speciated. Alternatively, 'association by descent' predicts that the lineage isolating species C in the host should be older than the lineage to the one from node 'a' to species A. To reconcile the expectation with the apparent reversal of the position for parasite 2 we have to postulate a sequence like that given in Fig. 4.2. The independent derivation of parasite 1, and the ancestors of 2 and 3 from a similar origin requires that parasite 1 now found on A must have either failed to parasitize B and C, or have become extinct upon them. Similarly, to account for a common ancestry of parasites 2 and 3 requires a similar failure of parasitism or extinction of a parasite on A. The net result is that there are now six speciation events, instead of three, required to explain the present-day pattern of evolution as 'association by descent'.

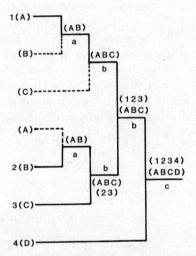

Fig. 4.2. Hypothetical history for the parasite group given in Fig. 4.1. (b). Dotted lines represent lineages failing to be parasitized, or losing parasites (extinction) — the hosts, parasites, and their ancestors are represented at the tips and nodes as upper case letters and numbers and equivalent components in lower case letters.

Interpreting cladograms as successive speciation events at the nodes makes predictions about the sequence or history of parasitism that is based on a character analysis of the parasites themselves. As Mitter and Brooks (1983) point out there is an obvious resemblance between 'association by descent' versus 'colonization' in historical parasitism with 'vicariance' versus 'dispersal' events in historical biogeography. Parasites occur in hosts either because they speciated at the same time for the same reasons or because they invaded from another source. Geographical disjunctions between related taxa showing similar patterns to unrelated taxa are hypothesized as occurring simultaneously and a result of historical, geographical, or ecological events. Non-correspondence between groups occurring in the same areas invites *ad hoc* explanations of which dispersals and extinctions are two.

3. *Correspondence between host and parasite*

When comparing host with parasite phylogenies it is difficult to test any hypothesis other than one of 'association by descent'. Hypotheses based on extinctions, invasion by colonization, and also host switching, all require *ad hoc* explanations because they are not based on any empirical scheme of either the host or parasite relationships. Hypotheses based on 'association by descent' predict the least number of character changes or speciation events and thus can easily be tested by new data (Mitter and Brooks 1983). In the examples to be considered here, there is a degree of discordance in the phylogenetic hypotheses of hosts when compared to those of the parasites. This is for various reasons. For example, there can be several, equally parsimonious, phylogenetic solutions for a given set of data. Secondly, there are overlapping parasite distributions, in *Nothofagus betuloides* for example (see p. 66), which have phylogenetic connections in different directions. Thirdly, not all host species within one monophyletic group are parasitized by an equal number of typo-pathogens. The method

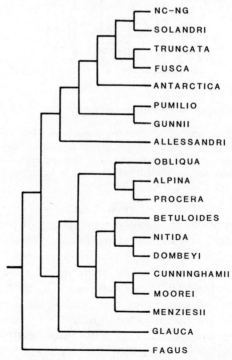

Fig. 4.3. *Nothofagus* cladogram (tree no. 40) rooted with the sister group *Fagus*. See text.

adopted here is to compare all of the available topologies generated by Swofford's PAUP program for the hosts with those of the parasites. The criterion of choice is that combination which maximized the number of common components and minimized the number of colonizations or extinctions under a hypothesis that coevolution occurs by 'association by descent'. Admittedly, this gives an *a priori* bias towards coevolution, but has the merit of choosing a result in terms of explanatory power.

Nothofagus: the host phylogeny

The genus *Nothofagus* comprises about 35 species (Humphries 1981, 1983) which occur in New Guinea, New Caledonia, New Zealand, Tasmania, southern and eastern Australia, and in Pacific South America in Chile and Argentina. Here, we concentrate on those 17 species in South America, Australia, Tasmania, and New Zealand which are included in the subsections *Antarcticae, Pumilae, Quadripartite*, and *Tripartite* of van Steenis (1953, 1971) (see Figs 4.3 and 4.4). The analyses presented here are new, based on 32 characters (e.g. Philipson and

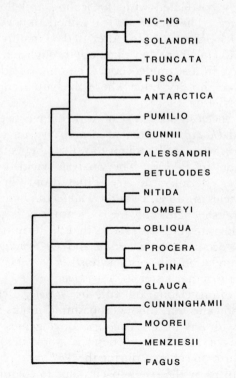

Fig. 4.4. Nelson consensus tree for all *Nothofagus* trees. See text for explanation.

Philipson 1979) and new interpretations of character transformations in the light of recent criticisms (Heads 1985; Humphries 1985). Onehundred-and-ninety equally parsimonious trees were generated with a consistency of 0.761. The cladogram which showed maximal fit with the parasite groups is given in Fig. 4.3, and to show that the topologies varied only because of zero branch lengths and alternative basal positions for plesiomorphic taxa, a Nelson consensus tree is given in Fig. 4.4 (see Nelson and Platnick 1981; Myamoto 1985).

The parasites

1. *Heterobathmia*

The genus *Heterobathmia* is the sole member group of the lepidopteran suborder Heterobathmiidna and is the sister group to all proboscis-bearing Lepidoptera, the Glossata (Kristensen and Nielsen 1983). Species of *Heterobathmia* are univoltine (i.e. with one generation per season), and the adults fly during late winter and early spring. The adult activity periods are closely correlated with the leafing and flowering of the deciduous *Nothofagus*. The larvae are leaf miners in newly set leaves of *Nothofagus* and development is rapid. After 10 days mature larvae burrow into the soil to 8-15 cm where they diapause through summer, autumn, and most of winter in a strong cocoon. The pharate adults tunnel up through the soil to the dry litter where the pupal cuticles are shed (Kristensen and Nielsen 1983).

There are 10 species, only two of which have been named, *H. pseuderiocrania* and *H. diffusa* (Kristensen and Nielsen 1979).

The data matrix was small enough to utilize the exhaustive branch and bound option in PAUP. The analysis yielded three equally parsimonious trees. Because the trees differed only in the topologies relating to the species group C, F, and G, all feeding on *N. obliqua*, and with basal zero branch lengths only the consensus tree is given (Fig. 4.5).

A reduced cladogram (Fig. 4.6) shows that the terminal sister species pair '*A*' and '*B*' occur in *N. antarctica* and *N. pumilio*, respectively, implying that the relationships of their hosts is *N. antarctica* sister to *N. pumilio* and those together sister to *N. obliqua*.

A comparison of this cladogram with the *Nothofagus* cladograms and consensus tree indicates that the heterobathmid moths show no direct correspondence with any of the *Nothofagus* phylogenies. At the species level an hypothesis of 'association by descent' would involve many more extinctions or failures to colonize during the *Nothofagus* phylogeny. More likely is opportunism by the sister species pair to colonize *N. antarctica* and *N. pumilio* from an origin on *N. obliqua*.

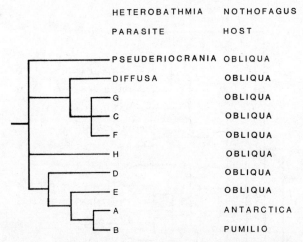

Fig. 4.5. Consensus tree for ten species of *Heterobathmia*.

Fig. 4.6. A comparison of the implied host relationship derived from the *Heterobathmia* consensus tree (see Fig. 4.5) with a simplified expression of the host cladogram (see Fig. 4.3).

2. Eriococcidae

The Eriococcidae are a family of scale insects (Homoptera: Coccoidea), a group in which the largely sedentary females feed on the sap of host plants. Many scale insects cause damage to the host not only by depletion of sap, but also indirectly by the transmission of viruses, the injection of toxins, and contamination by honeydew. Although the species that have been placed in the Eriococcidae probably do not form a monophyletic group, certain genera, parts of genera, or groups of genera do appear to be monophyletic. The 22 species discussed here belong to the genera *Eriococcus* and *Madarococcus*. *Eriococcus* (all species of which may not form a monophyletic group) occurs throughout the world on a wide variety of hosts. *Madarococcus* was erected for a few New Zealand species, and differs from *Eriococcus* by only a single character. All species of the Eriococcidae with spatulate suranal setae belong to one of these two genera, and all appear to be confined to *Nothofagus*. The PAUP analysis yielded more than 100 cladograms with a consistency of 0.394. However, there were only three basic topologies varying in terms of the position of *Eriococcus rubrifagi*, *E. fagicorticis*, *E. rotundus*, and *E. maskelli*.

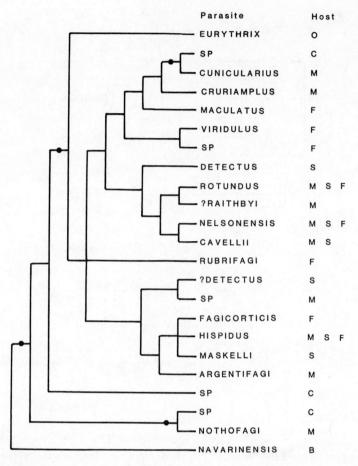

Fig. 4.7. Eriococcidae consensus tree for 22 species. B, *Nothofagus betuloides*; C, *N. cunninghamii*; F, *N. fusca*; M, *N. menziesii*; O, *N. obliqua*; S, *N. solandri*.

Because the different positions did not really affect the comparisons to *Nothofagus*, a Nelson consensus tree is given (Fig. 4.7).

A comparison of this consensus tree with the *Nothofagus* cladogram (Fig. 4.3) suggests that there are two groups of eriococcids with similar and fairly obvious correspondence with *Nothofagus cunninghamii* and *Nothofagus menziesii*, some general support for the relationship of *Eriococcus eurythrix* with New Zealand and Australian taxa, and generally the outgroup/sister group is South American (Fig. 4.8). However, at the species level the promiscuous distribution of *E. rotundus*, *E. nelsonensis* and *E. hispidus* on *Nothofagus fusca*, *N. menziesii* and *N. solandri*, together with the host-implied sister group relationships of *N. menziesii* with *N. solandri*, and *N. fusca* with *N. menziesii* and *N. cunninghamii*, suggest

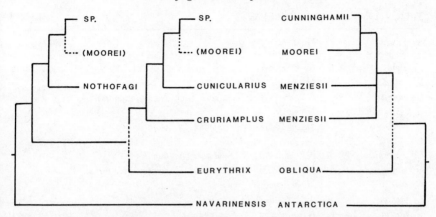

Fig. 4.8. Comparison of Eriococcidae consensus tree (see Fig. 4.7) with host *Nothofagus* cladogram (see Fig. 4.3). See text for explanation.

that, for New Zealand at least, an hypothesis of 'association by descent' would be highly unlikely. Nevertheless, the phylogenies of the two sister species pairs, *Eriococcus* sp. on *N. cunninghamii* and *E. nothofagi* on *N. menziesii*, together with *Eriococcus* sp. on *N. cunninghamii* and *Madarococcus cunicularius* on *N. menziesii* are too similar to the host phylogenies to be due to chance. Comparison to the *Nothofagus* cladograms shows that the sequence for *E eurythrix* on *N. obliqua* and *E. naverinensis*, the outgroup species, on *N. antarctica* follows a general biogeographical pattern (Fig. 4.8).

3. *Cyttaria*

Cyttaria is a genus of eleven gall-forming parasitic fungi known only from branches and smaller twigs of twelve southern temperate species of *Nothofagus*. The pattern of pathogenicity is complicated because some species occur on one host whilst others occur on two or more; *C. hariotii* occurs on five different hosts (Table 4.1). The taxonomy of the species has been well worked out (see Santesson 1945; Rawlings 1956; Kobayashi 1966; Gamundí 1971), but studies on species interrelationships are few (Rawlings 1956; Kobayashi 1966; Korf 1983). Rawlings (1956) divided the group into two — those species with globose stromata versus those with longitudinal stromata (see Fig. 4.13a). Korf (1983) advanced a phylogeny on the principle that 'fungi are accurate taxonomists'. He recognized three groups (see Fig. 4.13c) based on geographical position, host range, and the infrageneric groups of the host as suggested by pollen types.

To make an independent analysis we used 30 characters derived from the systematic papers cited above. Two analyses using PAUP were

Table 4.1. *Cyttaria* species and hosts

Species	Hosts
C. nigra	N. menziesii
C. pallida	N. menziesii
C. gunnii	N. menziesii, N. solandri (incl. N. cliffortioides)
C. septentrionalis	N. moorei
C. hookeri	N. antarctica, N. betuloides, N. pumilio
C. berteroi	N. glauca, N. obliqua
C. darwinii	N. antarctica, N. betuloides, N. pumilio
C. hariotii	N. antarctica, N. betuloides, N. dombeyi, N. nitida, N. pumilio
C. espinosae	N. obliqua, N. procera
C. jahowii	N. betuloides, N. dombeyi
C. exigua	N. betuloides

undertaken, the first scoring character 2 as longitudinal for *Cyttaria jahowii*, the second treating the same character in the same taxon as unknown. The chosen outgroup, Boedijnopezizeae (Korf 1983; pers comm.), seemed only to define *Cyttaria* as a natural group. The first analysis gave 19 cladograms with a consistency of 0.56, the second 57 cladograms with a consistency of 0.59. Every cladogram was compared with every *Nothofagus* cladogram. Those showing the greatest number of components and least number of proposed colonizations to make the remaining data fit are shown (Table 4.2, Figs 4.9 and 4.10).

Table 4.2. The number of components common to three *Cyttaria* cladograms and the *Nothofagus* cladogram (40) in Fig. 4.3.

Cyttaria cladogram (analysis in parenthesis)	Number of common components	Postulated colonizations	Number of colonizations
19 (1)	10	1–7	7 (Fig. 9)
12 (1)	7	1–7	7
40 (2)	9	1,2,7	3 (Fig. 10)

Colonizations: 1, *C. hookeri* on to *N. pumilio*; 2, *C. hookeri* on to *N. antarctica*; 3, *C. darwinii* on to *N. pumilio*; 4, *C. hariotii* on to *N. pumilio*; 5, *C. darwinii* on to *N. antarctica*; 6, *C. hariotii* on to *N. antarctica*; 7, *C. gunnii* on to *N. solandri*.

None of the cladograms showed a 100 per cent fit without postulating some parasitism by colonization. In Fig. 4.9, the *Cyttaria* cladogram (no. 19) from the first analysis shows 10 components in common with 7 of those in *Nothofagus*. However, to make such a fit it is necessary to postulate seven colonizations for four of the *Cyttaria* species; Fig. 4.10 is a cladogram which is the best fit. It minimizes the number of

Nothofagus and its parasites

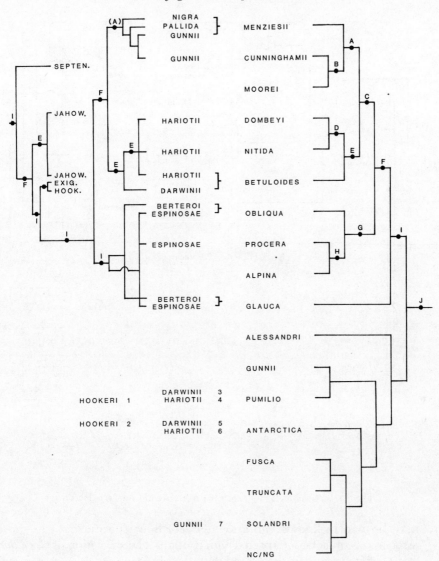

Fig. 4.9. *Cyttaria* cladogram (19) from first analysis (see text) superimposed onto the *Nothofagus* cladogram (Fig. 4.3).

colonizations to three, for two species, and shows only one less obvious component than the *Nothofagus* cladogram.

To have a complete hypothesis that accounts for all of the data in both the *Cyttaria* parasites and their hosts as represented in Fig. 4.10, a sequence resembling that in Fig. 4.11 would be required to account for the 'mismatches'. However, because the distribution of *Cyttaria* in the field

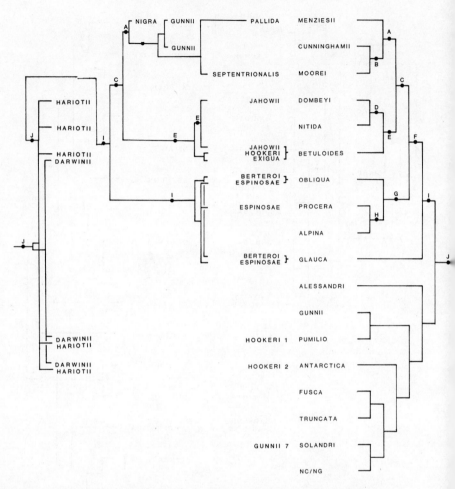

Fig. 4.10. *Cyttaria* cladogram (40) from second analysis (see text).

may be poorly known, absence of records of specific parasite–host associations must be interpreted with caution. The speciation of *C. pallida* and *C. septentrionalis* is associated with the phylogeny of *N. menziesii* and the ancestor of *N. cunninghamii* and *N. moorei*. If the *Nothofagus* phylogeny is representative then *C. septentrionalis* failed to invade *N. cunninghamii* in the speciation of this and *N. moorei*. *C. gunnii* is also associated with the speciation of *N. menziesii*, and the ancestor of *N. cunninghamii* and *N. moorei*, but failed to become a parasite of *N. moorei* in the speciation of *N. cunninghamii* and *N. moorei*. To account for the presence of *C. nigra* on *N. menziesii* by an 'association by descent' hypothesis, we have to postulate its presence at the time of speciation of *N. menziesii* and the ancestor of its sister species. Consequently, the only postulate that we

Nothofagus and its parasites

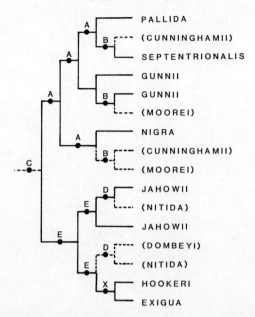

Fig. 4.11. Hypothetical history for *Cyttaria* based on the cladograms in Fig. 4.10. Dotted lines represent those lineages failing to become parasitized or losing parasites (extinction) from the hosts represented at the tips in parentheses. The components common to both host and parasite lineages are indicated by capital letters. X is a unique component of the fungal lineage.

can safely make is that *C. pallida*, *C. nigra*, *C. septentrionalis*, and *C. gunni* are all associated with the Australasian evergreen species of *Nothofagus*, but only one component 'A' can be ascribed to coevolution. In Fig. 4.11 we have also given a similar scheme for *C. jahowii*, *C. hookeri* and *C. exigua* on the South American species *Nothofagus betuloides*, *N. nitida* and *N. dombeyi*. Here it is important to note that the sister group relation of *C. hookeri* and *C. exigua* implies a speciation event (labelled component X) independent in origin from the phylogeny of *Nothofagus*. By working through the whole phylogeny it has been possible to identify all of the components which can be postulated to be coevolutionary events. In Fig. 4.12 it is clear, as indicated by component 'I', that *Cyttaria espinosa* and *C. berteroi* evolved at the same time as the ancestor for all of the *C. pallida*/*C. exigua* clade. To summarize the maximum number of coevolutionary events we have redrawn Fig. 4.12a in Fig. 4.12b. There are five component nodes common to both *Cyttaria* and *Nothofagus*, labelled A, C, E, I, and J. Component X is an event unique to the fungal group. To assess the extent to which coevolution has occurred we can postulate that speciation was wholly dichotomous. For 11 parasite species there are $n-1$ or 10 speciation events and thus 5 due to coevolution. Such

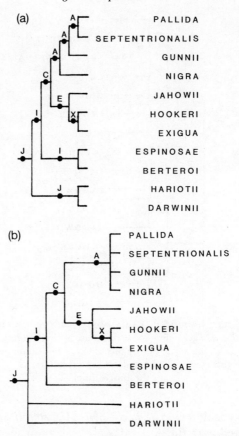

Fig. 4.12(a) Components common to both the *Cyttaria* cladogram and *Nothofagus* cladogram (see Fig. 4.10) expressed on the *Cyttaria* cladogram. (b) Components of cladogram 12a expressed as single events.

an hypothesis, presented in Fig. 4.12b, means that the ancestor of the *C. pallida/C. exigua* group, *C. espinosa*, *C. berteroi*, (component I), *C. hariotii*, and *C. darwinii* were all extant at the origin of *Nothofagus*.

A comparison to previous phylogenetic schemes (Fig. 4.13) shows that the hypotheses of Kobayashi and Korf are in agreement on the integrity of the *C. gunnii/C. pallida/C. nigra/C. septentrionalis* group, but vary as to the internal arrangements. The best fit is Kobayashi's phylogeny (Fig. 4.13b) where component 'B' fits the *Nothofagus* phylogeny (Fig. 4.3). Kobayashi's suggestion that at least *C. berteroi* is sister group to the Australasian taxa agrees with our scheme. By contrast, Korf's phylogeny (Fig. 4.13c) places the Australasian taxa as the isolated sister group to all of the remainder, the South American taxa.

Nothofagus and its parasites

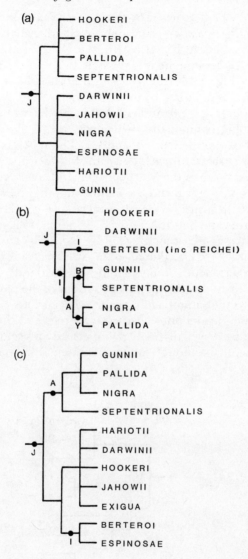

Fig. 4.13. Branching diagrams representing the different phylogenies of *Cyttaria* given by (a) Rawlings (1956), (b) Kobayashi (1966), (c) Korf (1983).

Historical biogeography

Nothofagus has been the subject of intense biogeographic study for the last 100 years or more because it is widely distributed across the southern Pacific Ocean (see Humphries 1981, 1985). By examining the components of the host and parasite phylogenies it should be possible to examine whether the trans-Pacific disjunctions are common to each of them. *Heterobathmia* occurs only in South America. By substituting areas for the taxa in Figs 4.3, 4.8, and 4.12b and then reducing them to single area statements it is possible to arrive at the area cladograms as given in Fig. 4.14. The *Nothofagus* hosts (Fig. 4.3) have seven area components, as indicated by the nodes in Fig. 4.14. Both the Eriococcidae and *Cyttaria* (Figs 4.8 and 4.12b) have the same two trans-Pacific biogeographic components which are identical to the host patterns (indicated by components A and C). The host taxa and the parasites show an Australasian track that runs between the central eastern part of Australia, north of the Macpherson–Macleay overlap and the southern mountains of Victoria and Tasmania, and into New Zealand, and then a trans-Pacific track to South America (Fig. 4.15). This is entirely separate from the South America/Tasmania track of *N. pumilio* and *N. gunnii* (which is not at all affected by parasites) and the New Zealand/ New Guinea/New Caledonia/South America track (which is affected by *Cyttaria* only in South America and New Zealand). It seems most likely that the present day disjunct distributions of the parasites and the hosts have shared the same Pacific history. Because *Cyttaria* has possibly two trans-Pacific disjunctions, *Nothofagus* three, and the Eriococcidae one, the groups were well diversified and already present in the early history of the Pacific. Although many commentators have attributed such

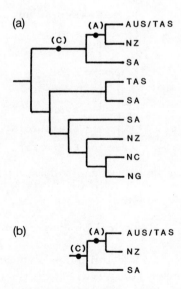

Fig. 4.14. Reduced area cladograms (a) *Nothofagus* (see Fig. 4.3) (b) Eriococcidae (Fig. 4.8) and *Cyttaria* (Fig. 4.12).

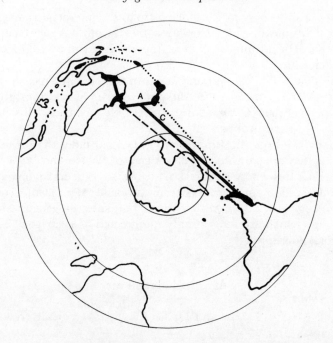

Fig. 4.15. Reduced area cladograms from Fig. 4.14a,b expressed as biogeographical tracks.

patterns to the break up of Gondwanaland (see Humphries 1981), Croizat (1964) for example, has repeatedly shown that the New Guinea/Eastern Australia New Caledonia/New Zealand track is a very distinct western Pacific component (see Humphries 1985). Our analyses have shown that in *Nothofagus* and its parasites this track can be subdivided into two sub-tracks linking eastern and southern Australia and Tasmania with New Zealand, and linking New Guinea and New Caledonia with New Zealand. Both of these link a more general Pacific component separately across the southern Pacific to South America. As this pattern is repeated by many groups (see e.g. Croizat 1964) it is most likely to be associated with vicariism in the Pacific rather than an alternative explanation. For references to the speculative geophysical theories that might account for trans-Pacific disjunctions see Humphries and Parenti (1985), and Nur and Ben Avraham (1981).

Conclusions

(1) To determine patterns of coevolution it is necessary to develop empirical methods that independently assess, but can be equally applied to hosts and parasites.

(2) It is possible only to examine host/parasite interactions in a testable theoretical framework of 'association by descent'. As yet there is no testable theoretical framework to examine parasitism by colonization or host-switching.

(3) Analysis of *Nothofagus* and some of its parasites has shown that coevolution only partially accounts for patterns of relationship and at best 50 per cent, in *Cyttaria* for example. Nevertheless, this is highly significant in such groups.

(4) Coevolved parasites may be as old, if not older than, the group with which they are parasitologically associated. Where they appear older, colonizations from now extinct hosts may have been involved.

(5) At least some components of coevolution are significantly correlated with biogeography. *Nothofagus* and its parasites show corresponding components which are most simply interpreted as vicariance events in Pacific biogeography.

Acknowledgements

We thank Mr O. Seberg and Dr G. Cassis for critically reading the manuscript.

References

Croizat, L. (1964). *Space, Time and Form; the Biological Synthesis*. Published by the author, Caracas.
Eichler, W. (1941). Wirtsspezifitat und stammesgeschichtliche Gleichlaufigkeit (Fahrenholzsche Regel) bei parasiten im allgemeinen und bei mallophagen imbesonderen. *Zool. Anz.* 132, 254–62.
Farris, J. S. (1970). Methods for computing Wagner trees. *Syst. Zool.* 19, 83–92.
Felsenstein, J. (1978). The number of evolutionary trees. *Syst. Zool.* 27, 27–33.
Gamundí, I. J. (1971). Las Cyttariales sudamericanas. *Darwiniana* 16, 461–511.
Heads, M. (1985). Biogeographic analysis of *Nothofagus* (Fagaceae). *Taxon*, 34, 474–80.
Hendy, M. D. and Penny, D. (1982). Branch and Bound algorithms to determine minimal evolutionary trees. *Math. Biosc.* 59, 277–90.
Hennig, W. (1966). *Phylogenetic Systematics*. University of Illinois Press, Urbana, Il.
Humphries, C. J. (1981). Biogeographical methods and the southern beeches (Fagaceae: Nothofagus) In *Advances in Cladistics: Proceedings of the First Meeting of the Willi Hennig Society* (eds V. A. Funk and D. R. Brooks), pp. 177–207. New York Botanical Garden.
—— (1983). Biogeographical explanations and the southern beeches. In *Evolution, Time and Space: the Emergence of the Biosphere* (eds R. W. Sims, J. H. Price, and P. E. S. Whalley), pp. 335–65. Academic Press, London.

—— (1985). Temperate biogeography and an intemperate botanist. *Taxon* 34, 480-92.
—— and Funk, V. A. (1984). Cladistic methodology. In *Current Concepts in Plant Taxonomy* (eds V. H. Heywood and D. M. Moore), pp. 323-62. Academic Press, London.
—— and Parenti, L. R. (1986). *Cladistic Biogeography*. Oxford monographs in biogeography, Clarendon Press, Oxford.
Kellogg, V. L. (1896). New Mallophaga, 1. *Proc. Calif. Acad. Sci.* 6.
Kluge, A. G. and Farris, J. S. (1969). Quantitative phyletics and the evolution of anurans. *Syst. Zool.* 18, 1-32.
Kobayashi, Y. (1966). On the genus *Cyttaria*. *Trans. mycol. Soc. Japan* 7, 118-34.
Korf, R. P. (1983). *Cyttaria* (Cyttariales): coevolution with *Nothofagus* and evolutionary relationship to the Boedijnopezizeae (Pezizales, Sarcoscyphaceae). *Aust. J. Bot.*, Suppl. ser. 10, 77-87.
Kristensen, N. P. and Nielsen, E. S. (1979). A new family of micropterigid moths from South America. A contribution to the morphology and phylogeny of the Micropterigidae, with a generic catalogue of the family Lepidoptera: Zeugloptera). *Steenstrupia* 5, 69-147.
—— —— (1983). The *Heterobathmia* life history elucidated: immature stages contradict assignment to suborder Zeugloptera (Insecta, Lepidoptera). *Zeit. zool. Syst. Evolforsch.* 21, 101-24.
Luckow, M. and Piaentel, P. A. (1985). An empirical comparison of numerical Wagner computer programs. *Cladistics* 1, 47-66.
Metcalf, M. M. (1920). Upon an important method of studying problems of relationship and of geographical distribution. *Proc. natn. Acad. Sci.* 6, 432-3.
—— (1929). Parasites and the aid they give in problems of taxonomy, geographical distribution and paleogeography. *Smithson. misc. Contr.*, 81, 1-36.
Mitter, C. and Brooks, D. R. (1983). Phylogenetic aspects of coevolution. In *Co-evolution* (eds D. J. Futuyma and M. Slatkin), pp. 65-98. Sinauer Associates, Sunderland, Massachusetts.
Myamoto, M. M. (1985). Consensus cladograms and general classifications. *Cladistics* 1, 186-9.
Nelson, G. and Platnick, N. I. (1981). *Systematics and Biogeography: Cladistics and Vicariance*. Colombia University Press, New York.
Nur, A. and Ben Avraham, Z. (1981). Lost Pacific continent: a mobilistic speculation. In *Vicariance Biogeography: a Critique* (eds G. Nelson and D. E. Rosen), pp. 341-58. Colombia University Press, New York.
Patterson, G. (1982). Morphological characters and homology. In *Problems of Phylogenetic Reconstruction* (eds K. A. Joysey and A. E. Friday), pp. 21-74. Academic Press, London.
Philipson, W. R. and Philipson, M. (1979). Leaf vernation in *Nothofagus*. *N.Z. J. Bot.* 17, 417-21.
Rawlings, G. B. (1956). Australasian Cyttariaceae. *Trans. R. Soc. N.Z.* 84, 19-28.
Santesson, R. (1945). *Cyttaria*, a genus of inoperculate discomycetes. *Svensk bot. Tidskr.* 39, 319-45.
Schlinger, E. I. (1974). Continental drift, *Nothofagus* and some ecologically associated insects. *Ann. Rev. Ent.* 19, 232-343.

Steenis, C. G. G. J., van (1953). Results of the Archbold Expeditions. Papuan *Nothofagus*. *J. Arnold Arbor.* 34, 301–74.
—— (1971). *Nothofagus*, key genus of plant geography, in time and space, living and fossil, ecology and phylogeny. *Blumea* 19, 65–98.
Swofford, D. (1984). *PAUP—Phylogenetic analysis using parsimony*. Illinois Natural History Survey, Champaign. Unpublished program and users manual.

5. Coevolutionary relationships of lice and their hosts: a test of Fahrenholz's Rule

C. H. C. LYAL

Department of Entomology, British Museum (Natural History), U.K.

Summary

The traditional view of louse phylogeny as being governed by phyletic tracking of the hosts (Fahrenholz's Rule) underpins much of louse systematics. This reliance on host relationships as indicators of louse relationships prevents rather than aids studies of coevolution. To determine the full implications of using Fahrenholz's Rule predictions of the phyletic tracking model are tested against observations of louse specificity and host relationships, considering especially a cladogram of 350 louse species derived independently of host information. Predictions of strict host cospeciation and failure of lice to colonize novel hosts are shown to be falsified in a number of cases, and the model fails to explain a minimum of 20.7 per cent of louse speciation events in the history of the 350 species analyzed. The resource tracking model, sometimes suggested as an alternative to the phyletic tracking model, is considered as the other extreme of a continuum of host-parasite relationships. The position of any parasite on this continuum is governed by a number of factors, which are discussed. In view of its failure to explain all host-parasite associations, Fahrenholz's Rule can have no value as a precise tool in phylogenetic reconstruction for lice and their hosts.

©The Systematics Association, 1986. This chapter is from *Coevolution and systematics* (eds A. R. Stone and D. L. Hawksworth) published for the Systematics Association by the Clarendon Press, Oxford.

Introduction

The lice (Phthiraptera) are a group of obligate permanent ectoparasites of birds and mammals, lacking any free-living stage. The majority of lice are known as parasites of single host species. The high level of monoxenia and the clear dependence of the lice on the dermecos of the hosts might be expected to lead to a high level of coevolution of lice and their hosts, itself resulting in phylogenetic concordance between the two. This assumption has been formalized in Fahrenholz's Rule, which states that the phylogenies of host and parasite are topologically identical (because of coevolution and cospeciation). This rule (and its unwritten precursors) have been accepted as axiomatic by many taxonomists and systematists of lice for at least the last 50 years, and classifications proposed for lice have been strongly influenced by the classifications of the hosts. The corresponding view, that host relationships will be indicated by the relationships of their parasites, is also held, and proposals for the reclassification of hosts have been made on the basis of louse relationships (e.g. Webb 1949; Timmermann 1957; Kettle 1977; Timm 1983). In some cases this argument has been reduced to complete circularity, the postulated relationship of the hosts being used to infer louse relationship, which in turn is used to support the initial hypothesis of host relationship (Traub 1980). Although some systematists do recognize that louse relationships do not always mirror host relationships (Eveleigh and Amano 1977), there is still a strong belief in the general application of Fahrenholz's Rule, and a less than critical appreciation of how its action can be demonstrated (Timm 1983). The present situation is that Fahrenholz's Rule is questioned only in special cases, and otherwise is assumed to govern louse–host associations. The continued use of the undefined term 'relationship' reflects a lack of rigour in analytical techniques, without which no critical evaluation of the governance of Fahrenholz's Rule can be made. In the absence of such evaluation there can be no sound basis for acceptance of or justification for use of parasitological evidence in host classification, or host evidence in parasite classification.

Fahrenholz's Rule is primarily a hypothesis of ecological relationships through time, as manifested in coevolutionary phylogenetic relationships. As such it is open to test in three ways: (1) by examination of present-day ecological relationships of parasite and host to see if they conform to the predictions made necessary by the operation of Fahrenholz's Rule; (2) by examination of present-day host–parasite associations to determine whether they are consistent with the operation of Fahrenholz's Rule; (3) by comparison of host and parasite phylogenies to see if they are congruent, as predicted by Fahrenholz's Rule.

The objectives of this paper are: (1) to test the effectiveness of

Fahrenholz's Rule in describing louse–host coevolutionary relationships; (2) to indicate alternative hypotheses of louse–host coevolutionary relationships; and (3) to discuss the implications for systematic work on both lice and their hosts.

Effectiveness of Fahrenholz's Rule

In this section the various predictions made necessary by the operation of Fahrenholz's Rule will be listed and discussed briefly. Most formulations of Fahrenholz's Rule are very precise, and this permits very rigorous testing. Such precision of the Rule is necessary if it is to be used as a systematic tool. The more generalized format 'Related hosts tend to have related parasites' ('Clay's Rule') is not predictive in detail, and cannot be tested rigorously. This latter format is useful in parasite

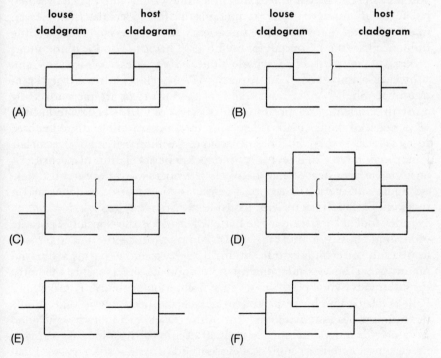

Fig. 5.1. (A–F). Comparisons of louse and host cladograms to show adherence to and departures from coevolution as expressed by Fahrenholz's Rule. (A) Adherence to Fahrenholz's Rule; (B) Independent speciation of the lice; (C) Independent speciation of the hosts; (D) Speciation of hosts and lice subsequent to independent host speciation: association of holophyletic and paraphyletic groups; (E) Independent speciation of the hosts or loss of lice: association of holophyletic and paraphyletic groups; (F) Secondary infestation: association of holophyletic and polyphyletic groups.

identification work (given the present paucity of good keys to lice), but not in systematic studies.

The first part of the hypothesis embodied in Fahrenholz's Rule to be considered is that of strict cospeciation. For the phylogenies of host and parasite to be topologically identical (Fig. 5.1A) no speciation (cladogenesis) of one partner can take place without speciation of the other. In the absence of other complicating factors (which themselves would be denied under Fahrenholz's Rule), this hypothesis provides the following testable predictions: (1) the number of parasite species will be the same as the number of host species; (2) no species of host (in any given host clade) will be associated with more than one species of parasite (of a parasite clade that had its origin later than the host clade); and (3) no species of parasite will parasitize more than one species of host.

In order to test these predictions it must be assumed that the taxonomy of both host and parasite species is 'correct', and that species of host and parasite can be compared meaningfully. One group of lice in which this is not an unreasonable assumption is the Trichodectidae (ischnoceran mammal lice), in which the alpha-level taxonomy is relatively little influenced by the taxonomy of the hosts. This is not so for many other groups of lice. In the Trichodectidae 337 species (350 species and subspecies) parasitize 244 species (577 species and subspecies) of mammal. Slightly over 34 per cent of the host taxa are parasitized by more than one taxon (species or subspecies) of trichodectid, and about 18 per cent of the trichodectid species are found on more than one host species (Lyal, in prep. *a*). These imbalances, whilst falsifying Fahrenholz's Rule, may have come about by independent speciation of parasite or host or by transfer of lice between members of different host taxa (secondary infestation). Secondary infestation is discussed in detail below, independent speciation being considered first.

Independent speciation of the louse will lead to sister-species of parasite on a single host species (Fig. 5.1B). Although this situation may arise as a result of secondary infestation, if the two louse species are found on no other hosts, independent speciation of the louse is the most parsimonious hypothesis. Should such a situation be found, Fahrenholz's Rule is falsified in this respect. The lack of cladistic analysis in lice limits the number of cases available, but in the Trichodectidae there is good evidence for independent speciation of lice on the host having occurred seven times, and more equivocal evidence for a further seven cases (Lyal, in prep. *a*). This is just over 7 per cent of the speciation events in the analyzed history of the family and, as only the more recent events can be reconstructed in this way, it is probably an underestimate.

Independent speciation of the host will lead to a single parasite species associated with sister-species of host (Fig. 5.1C) or one of these sister-species not being parasitized (Fig. 5.1E). Cases of the first type are virtually impossible to detect in the absence of a cladistic analysis of the hosts, although there are a number of cases of host congeners sharing the same species of lice. Independent speciation of the host may be responsible for such situations, but they may also have arisen following transfer of lice between hosts. Such transfer may be considered especially likely to occur between closely related hosts (see below). Although the possibility of distinguishing the origin of the situation is of interest here, it should be noted that either explanation involves a falsification of Fahrenholz's Rule. The second possibility—absence of a parasite from one of a pair of sister-species—is also difficult to recognize. Without cladistic analysis of the hosts one must look to absence of a parasite from one of a recognizable host group that can be inferred to have had an ancestral association with the parasite taxon. Unfortunately, there are alternative explanations: lice may be present, but not yet collected, or they may have been present, but have become extinct. (It is notable that extinction is also a falsifier of Fahrenholz's Rule.) To avoid these difficulties, at least partially, an adequate test can be made by examination of the present distribution of a single louse taxon in relation to host distribution. As speciation of the host can theoretically take place in any part of its range, the parasite of that host must be present in all parts of that range if cospeciation is to occur. Although this is still difficult to test, both because of inadequate collecting and the sometimes-expressed belief that the presence of the host must indicate the presence of the parasite (e.g. Kloet and Hincks 1964), there are recorded instances where host populations lack lice (Plomley and Thompson 1937; Thompson 1940; Hopkins 1945, 1949; Clay 1949, 1964, 1966, 1976; Boyd 1951; Klockenhoff 1972; Ledger 1980; Timm and Price 1980; Timm 1983). This 'secondary absence' can lead to a situation in which Fahrenholz's Rule does not apply and is therefore falsified.

The second part of the hypothesis embodied in Fahrenholz's Rule is that secondary infestation does not take place. Clearly, no hypothesis of genealogical relationships of one partner can be based on those of the other partner if the two may not have been in association for all of the period under consideration. A few instances of secondary infestation have been recognized for many years, but such cases have been believed to be very infrequent (Hopkins 1949, 1957), confined to a very early stage in phthirapteran evolution (Clay 1949, 1957), or subject to phyletic constraints (Traub 1980). This belief in the rarity and limited nature of secondary infestation, and the consequent lack of challenge to Fahrenholz's Rule, has been maintained by the apparent feeling that all secondary infestation must be recognizable immediately. Thus, the cases admitted in the literature are mostly of transfer between very distantly related host taxa—Falconiformes and Columbiformes (Price and Beer 1963), Hystrichomorphia and Lagomorphia (Hopkins 1949), wallaby and dog (Hopkins 1949, 1957; Kéler 1971), Aves and Mammalia (Tarry 1967; Clay 1970, 1971). In most considerations of louse groups, where a (presumed) monophyletic group of lice is restricted to a (presumed) monophyletic group of hosts, Fahrenholz's Rule is assumed to have been followed, and any minor discontinuities noticed are attributed to poor collection or simply ignored. If one of the factors acting to limit secondary infestation is the difference in environments available to parasites on different species of the host class, this difference might in many host groups be broadly linked to host (phyletic) relationships. If this is so, secondary infestation may be more common between closely related host animals than between those more distantly related, and thus the coincidence of monophyletic host and parasite groups does not indicate lack of secondary infestation. Careful analysis of louse–host associations in the Trichodectidae (Lyal in prep., a) reveals 16 extant cases of multiple host association resulting from host transfer, involving eight louse species. Of these 16 cases only three can be seen clearly to have taken place without human intervention in bringing the hosts together in unnatural proximity. The number of species examined to give these cases was 187, the proportion thus being 2.1 per cent (unequivocally natural) or 8.6 per cent (total). This proportion might be higher were the 102 species and subspecies of lice described from pocket gophers included, but these (and their hosts) have not been analyzed cladistically and the data are therefore not available. This association is discussed further below and reasons given for inferring a higher rate of secondary infestation.

With such a low figure for natural secondary infestation in Trichodectidae, the position of traditional louse systematics might seem tenable. However, there is a further test of the rate of secondary infestation by direct comparison of louse and host phylogenies, and this should be made before final conclusions are drawn. A cladogram of the louse family Trichodectidae is produced by Lyal (1985). This cladogram is 'converted' to a phylogenetic tree by postulating a one-to-one topological relationship between dichotomies on the cladogram and the branching of the tree, and making no a priori hypothesis about the topology of the phylogenetic tree when there are multiple furcations on the cladogram (Wiley 1979, 1981; see Platnick 1977, for a dissenting argument). The phylogeny thus produced can be compared to phylogenetic hypotheses of the hosts taken from the literature, and primary host associations of the clades of lice determined by parsimony methods (Lyal in prep. c). In many cases the distribution of host associations on the cladogram of lice is so remarkable that ancestral secondary infestations are apparent on first inspection. The louse cladogram depicts 198 cladogeneses (speciation events); the other cladogeneses in the history of the family (primarily lice parasitic on the large and systematically confused groups pocket gophers and hyraxes) are not resolved. Of the 198 speciation events a minimum of 41 (20.7 per cent) are found to be associated with secondary infestation. This figure is likely to be an underestimate, as lice parasitizing closely related hosts whose cladistic history has not been determined are assumed to have developed that association under the provisions of Fahrenholz's Rule (the null hypothesis being that there has been no independent evolutionary activity of the lice).

The figure of 20.7 per cent is plainly more significant than the 2 per cent determined for secondary infestation among living species. Two reasons can be found for this disparity: inadequate records and a high rate of parasite evolution. Records are inadequate because some phthirapterists dismiss records of lice from the 'wrong' hosts as 'stragglers' without looking for biological significance in their presence, and in some cases discard the specimens. The second factor is probably more significant. Price (1977) gives cogent reasons for believing evolutionary and speciation rates of parasites to be very high compared to those of their hosts. The operation of founder effect and genetic drift could cause colonizing lice isolated on a novel host to diverge from the parent population and speciate in a comparatively short time. If this does happen, relatively more secondary infestations will be detectable in the phylogenetic history of a clade than can be observed in extant species.

In summary, all of the predictions made by the operation of Fahrenholz's Rule can be and are falsified, indicating that for lice—both bird-infesting and mammal-infesting—Fahrenholz's Rule does not provide a full explanation for modern host–parasite associations.

Alternative hypotheses

The falsification of Fahrenholz's Rule does not mean that louse–host coevolution is never described adequately by the Rule, merely that other factors must be considered.

A model that apparently describes the association of some ectoparasites and their hosts rather more accurately than Fahrenholz's Rule was proposed by Kethley and Johnston (1975). This is resource tracking, and postulates that parasites are adapted to resources that are, or may be, distributed independently of the taxonomic relationships of the hosts. Kethley and Johnston demonstrated this for the highly host-specific syringophilid mites, which live inside the quills of birds. The relationships of the higher mite taxa are not congruent with those of the host taxa, and the authors suggested that the mites are distributed according to quill size, which itself is not distributed in a manner congruent with host relationship. It is notable that Kethley and Johnston used phenetic grouping methods rather than phyletic, and the hypothesis should be tested using phyletic relationships of mites and birds to be compared fully to the hypothesis of Fahrenholz's Rule.

The terms resource tracking and phyletic tracking (which is sometimes used to describe adherence to Fahrenholz's Rule) might be taken to imply some 'purpose' behind parasite distribution. This is not so, of course, as the terms refer to descriptions of the distributional results of process rather than the driving force behind that process.

Timm (1983) suggested that Fahrenholz's Rule and resource tracking are ends of a continuum of dispersal opportunity and niche availability, and he is undoubtedly correct in this. However, no group of parasites has yet been identified that falls at the extremes of this continuum, although some clearly approach them. It is therefore valuable to consider the factors operating that govern the position of a given parasite on the continuum. These are the same as those governing parasite distribution over host taxa and fall broadly into two classes: factors affecting colonization (i.e. governing the probabilities of a parasite transferring between two animals of the host class) and factors affecting establishment (i.e. governing the probability of the parasite being able to survive and reproduce on a colonized host).

A major factor related to the possibility of a parasite moving between hosts is the degree to which a parasite is necessarily restricted to the host (throughout the life of the parasite). Lyal (in prep. *a*) examines nine families of insect ectoparasites of vertebrates (two Phthiraptera, one Siphonaptera, two Heteroptera, and four Diptera), and finds a direct relationship between this distributional dependence of the parasite on the host (throughout the life of the parasite) and the percentage of monoxenous species in each family. Consideration of the probable ancestry of these families reveals that the restriction on independent dispersal occurred in at least some of them concomitant with the development of obligate parasitism and not as a result of host specificity. This factor, then, is of importance in establishing the likelihood of colonization of novel host taxa within any given parasite group, and explains a great deal of the variation in host specificity between groups.

Any given parasite type has a colonizing range, i.e. the distance over which an individual can detect and colonize an animal of the host class. For a strong-flying vagile insect like a hippoboscid fly this distance is far greater than that for a wingless louse, but within most parasite groups it is likely to be effectively constant. There is likely to be an inverse relationship between colonizing range of the parasite and the level of host specificity (Lyal in prep. *b*). Generally, the greater the degree of restriction to the host the lower the vagility, and thus the lower the colonizing distance and the fewer taxa in the host class encountered.

Furthermore, the more time spent on the host the greater the dependence on the host itself encountering other animals of the host class.

A hypothetical 'proximity index' can be envisioned for hosts of any given parasite type. For any pair of species in the host class, this is the number of encounters in unit time, an 'encounter' being an approach to within the colonizing distance of the parasite. The proximity index of any pair of species (or populations) in the host class will determine in part the probability of a parasite colonizing one of the pair from the other, this part increasing with the necessary distributional dependence of the parasite on the host. The proximity index will in most cases be higher between members of the same species than members of different species, and thus intra-specific passage of parasites will be more frequent than inter-specific passage. This is probably invariably the case with lice. The proximity index is dependent on overlap of ecotype, geographical range, population size, density of individuals, and temporal range. Each of these, then, has its effect on the colonizing ability of the parasite.

The colonizing distance for lice is probably in most cases very limited, colonization only taking place when two animals of the host class are in physical contact. However, it may be extended by phoresy, or by transfer on part of the environment (dust baths, burrow walls, detached feathers, etc.) (Lyal in prep. *b*). This renders the parameters of the proximity index for hosts of louse species slightly 'fuzzy'.

The proximity index does not give the whole picture. Lice, as demonstrated above and by Lyal (in prep. *a*), do not necessarily occupy the whole of the host range, so the proximity index should be calculated for only those populations of the host upon which the parasite is found. The greater the range of the host, the more other species of the host class with which it is likely to have a proximity index value above zero, and the more opportunities there will be for parasite colonization.

The final factor is the means of detection used by the parasite. A member of the host class can only be colonized by a parasite if the parasite recognizes it as such. Many stimuli are used by parasites to detect hosts (Marshall 1981), and these vary in specificity. It is clear that odours, for example (which are used by many parasite groups) may be far more specific than sight (which is used by the Hippoboscidae).

Having dealt with the factors affecting colonization, there remain those affecting establishment. The more time spent by a species in a given environment, the greater its adaptation to that environment is likely to be. Thus, the more dependent a parasite is on its host for distribution (see above), the more time it spends on the host, and the more closely adapted it is to the environment provided by the host. However, it cannot be argued that the adaptations of an animal to one environment preclude it from surviving in any other. If this were to be so, the original adoption of the ectoparasitic way of life could not have happened, and the phenomenon of secondary infestation could not occur. On the other hand, experiments involving transfer of lice from one host species to another frequently result in the death of the lice (Ewing 1933; Wilson 1934; Eichler 1936; Krynski *et al.* 1952; Ash 1960). Successful transfer, as it does occur (Eichler 1940; Tarry 1967; Lyal in prep. *a*) must involve some 'overlap' between the environmental tolerance of the parasite and the environmental conditions provided by the host. In most cases the environmental parameters important to the parasite are not known, and this is the case for all lice. The quill parameters described by Kethley and Johnston (1975) that apparently determine the success of syringophilid mites in establishing populations on birds are an outstanding exception to this general ignorance. Clearly, the less specific the parameters, the more the parasite will tend towards the resource tracking end of the spectrum. Some parameters will be characteristic of supra-specific taxa (like quills on birds), and so resource tracking can take place within taxonomic groups. The more homogeneous the group, and the less well-known the cladistic history of louse and host, the more difficult it is to determine the position on the continuum between resource tracking and phyletic tracking.

The last-mentioned situation is that found in the association between trichodectid lice of the genus *Geomydoecus* and rodents of the family Geomyidae (pocket gophers). Timm (1983) believed evolution within this association to be consistent with Fahrenholz's Rule, and used phenetically-based relationships of the lice to support postulated relationships of the hosts. However, the system is inconsistent with Fahrenholz's Rule in that the numbers of louse and host taxa differ, both at specific and subspecific levels, some louse species and subspecies are found on more than one taxon of host, louse and host ranges are not always congruent, and some species of gopher are parasitized by more than one species of *Geomydoecus*. A distribution map of lice and hosts of the *Geomys bursarius* complex of species (Timm and Price 1980; Timm 1983) shows eight species of lice and four species (eleven subspecies) of pocket gopher. Some of the lice are restricted to a distribution within the range of single subspecies or species, some are found on only one subspecies of a species, some are found on most subspecies of a species, some are found on more than one species, and one is found on all subspecies of one species and one subspecies of another in part of its range. The lice, in fact, have a geographical distribution superimposed upon, but partially independent from, the distribution of the gophers. The associations within the system are not consistent with the governance of Fahrenholz's Rule, and not all the anomalies can be accounted for by extinction and independent speciation. Resource tracking must therefore be invoked. This being the case, six assumptions are made to try and explain the system.

(1) Not all, if any, of the environmental parameters required by *Geomydoecus* species are restricted to individual rodent species.
(2) Populations of Geomyidae are analogous to islands in the MacArthur and Wilson (1967) model of island biogeography, and lice can colonize and establish populations upon these 'islands' in accordance with the model.
(3) The proximity index of pairs of populations of gophers is lower for populations of different taxa than it is for populations of the same taxon.
(4) The limitation of one taxon of lice per gopher population is due to an interactive process between the lice rather than any extrinsic factor.
(5) Populations of gopher lice cannot survive without gophers.
(6) Any geographical feature that prohibits gene flow between gopher populations, or provides a barrier limiting the distribution of gopher taxa, is likely to have the same effect on the lice of that gopher population or taxon.

The host and geographical distributions of gopher lice depicted by Timm and Price (1980), and Timm (1983) are entirely consistent with a system governed by the operation of these assumptions.

Systematic implications

It is clear from the discussion above that the invocation of Fahrenholz's Rule to make systematic hypotheses of parasite or host relationships carries with it a number of very precise predictions of host–parasite associations, and it is equally clear from study of the literature that these are generally ignored. Hennig (1966) discussed the misleading assumptions that can be made if these predictions are not considered in detail, but, as the majority of louse systematists are not cladists, this discussion has not been well received (or even noticed). A brief outline of the major points is repeated here.

Independent speciation of the parasite on the host, although falsifying Fahrenholz's Rule, does not inevitably lead to systematic problems, as in no case is a host (or holophyletic group of hosts) not associated with a holophyletic group of lice if Fahrenholz's Rule is otherwise adhered to (Fig. 5.1B). Independent speciation of the host does not immediately lead to systematic problems (Fig. 5.1C), but subsequent speciation of the parasite would cause a paraphyletic group of hosts to be associated with a holophyletic parasite group (Fig. 5.1D). If one of a pair of host sister-species is without an ancestral parasite, it will be seen to be excluded from a host group characterized by the possession of that parasite, and again a holophyletic group of parasites will be associated with a paraphyletic group of hosts (Fig. 5.1E). With sufficient secondary absences an association between a holophyletic parasite group and a polyphyletic host group is possible. Secondary infestation can lead to holophyletic groups of lice being associated with paraphyletic or polyphyletic host groups (Fig. 5.1F). In such cases postulation of holophyly of 'groups' of lice associated with holophyletic host groups within the assemblages are likely to be incorrect.

A major problem in the systematic application of Fahrenholz's Rule in almost all cases is lack of clarity in the category of 'relationship' used. To an evolutionary systematist the distinction made above between paraphyletic and holophyletic groups would perhaps be less important, although as Fahrenholz's Rule deals explicitly with cladistic groups this should not be the case. Most louse workers have not stated the type of groups they think they are dealing with, many perhaps believing that their groups are 'natural' and need no other description. Others appear to operate in the belief that taxa related by phenetic analysis are also related phylogenetically, although they do not state on what basis this assumption is made, or whether their phenetic groups are believed to be paraphyletic or holophyletic. Discussions on host relationships appear to indicate that the groups are believed to be holophyletic (e.g. Timm 1983), but the method of constructing a phenetic group so that it is holophyletic is not explained. This problem of clarity can be resolved by use of the cladistic system of analysis, and, indeed, no other analytical technique produces results which can be discussed in terms of Fahrenholz's Rule or other coevolutionary hypotheses.

The indication that Fahrenholz's Rule does not apply in all cases, and that departures from the coevolutionary processes postulated by Fahrenholz's Rule will produce erroneous hypotheses of relationships, sharply limits the application of Fahrenholz's Rule as as systematic tool. Clearly, if no set of associations can be discussed without full cladistic analysis of both parasite and host, use of Fahrenholz's Rule to predict one of the sets of cladistic relationships from the other is superfluous. This problem has been recognized in 'Hopkins' Principle', which equates the chances of a postulated host relationship being correct with the number of parasitic groups supporting that relationship (such groups could be species or holophyletic groups, although Hopkins did not present his ideas in cladistic terms). This may be a valid technique, providing all other methods of determining host relationships are unusable, if the following precautions are taken: (i) the phylogenies of all the parasites concerned are produced using cladistic techniques (Brooks 1981); (ii) a consistent set of host relationship hypotheses is generated by all or most of the parasite cladograms; (iii) the hypothesis of Fahrenholz's Rule is the most parsimonious explanation for the observed host–parasite associations, and no non-phylogenetic factor (e.g. high proximity index between hosts as a result of sympatry) provides a more or equally parsimonious explanation. Such a process is not a short-cut to finding host relationships, nor can it be used to establish louse relationships on the basis of those of their hosts.

References

Ash, J. S. (1960). A study of the Mallophaga of birds with particular reference to their ecology. *Ibis* 102, 93–110.
Boyd, E. M. (1951). A survey of parasitism of the starling *Sturnus vulgaris* L. in North America. *J. Parasit.* 37, 56–84.
Brooks, D. R. (1981). Hennig's parasitological method: a proposed solution. *Syst. Zool.* 30, 229–49.
Clay, T. (1949). Some problems in the evolution of a group of ectoparasites. *Evolution, Lancaster, Pa.* 3, 279–99.
—— (1957). The Mallophaga of birds. In *First Symposium on Host Specificity among Parasites of Vertebrates, Neuchatel*, pp. 120–55. University of Neuchatel.
—— (1964). Geographical distribution of the Mallophaga (Insecta). *Bull. Br. Orn. Club* 84, 14–6.
—— (1966). The species of *Strigiphilus* (Mallophaga) parasitic on the barn owls (*Tyto*). *J. Entomolog. Soc. Queensland* 5, 10–7.
—— (1970). The Amblycera (Phthiraptera: Insecta). *Bull. Br. Mus. Nat. Hist. (Ent.)* 25, 75–98.
—— (1971) A new genus and two new species of Boopidae (Phthiraptera: Amblycera). *Pacif. Insects* 13, 519–29.
—— (1976). Geographical distribution of the avian lice (Phthiraptera): a review. *J. Bombay Nat. Hist. Soc.* 71, 536–47.
Eichler, W. (1936). Die Biologie der Federlinge. *J. Orn. Lpz.* 84, 471–505.
—— (1940). Wirtsspezifität und stammegesgeschtlighte Gleichläufigkeit (Fahrenholzsche Regel) bei Parasiten im allgemeinen und bei Mallophagen im besonderen. *Zool. Anz.* 132, 254–62.
Eveleigh, E. S. and Amano, H. (1977). A numerical taxonomic study of the Mallophagan genera *Cummingsiella* (= *Quadraceps*), *Saemundssonia* (Ischnocera: Philopteridae), and *Austromenopon* (Amblycera: Menoponidae) from alcids (Aves: Charadriiformes) of the northwest Atlantic with reference to host–parasite relationships. *Can. J. Zool.* 55, 1788–801.
Ewing, H. E. (1933). *A Manual of External Parasites.* Charles C. Thomas, Springfield and New york.
Hennig, W. (1966). *Phylogenetic Systematics.* Translated by D. D. Davis and R. Zangerl. University of Illinois Press, Urbana.
Hopkins, G. H. E. (1945). Lice of hyraxes, especially *Procavia capensis. J. ent. Soc. S. Afr.* 8, 1–12.
—— (1949). The host-associations of the lice of mammals. *Proc. zool. Soc. Lond.* 119, 387–604.
—— (1957) The distribution of Phthiraptera on mammals. In *First Symposium on Host Specificity among Parasites of Vertebrates, Neuchatel*, pp. 88–119. University of Neuchatel.

Kéler, S. von (1971). A revision of the Australasian Boopiidae (Insecta: Phthiraptera) with notes on the Trimenoponidae. *Aust. J. Zool.* Suppl. 6, 1-126.

Kethley, J. B. and Johnston, D. E. (1975). Resource tracking in bird and mammal ectoparasites. *Misc. Publ. ent. Soc. Am.* 9, 231-6.

Kettle, P. R. (1977). *A Study on Phthiraptera (chiefly Amblycera and Ischnocera) with particular reference to the evolution and host parasite relationships of the Order*. Ph.D. Thesis, University of London.

Klockenhoff, H. (1972). Zur Taxonomie der auf der Salvadorikrähe *Corvus orru* lebenden Mallophaga—Gattung *Myrsidea* Waterston, 1915. *Bonn. zool. Beitr.* 23, 253-66.

Kloet, G. S. and Hincks, W. D. (1964). A check list of British Insects Part 1: Small orders and Hemiptera. *Handbk Ident. Br. Insects* 11, i-xv, 1-119.

Krynski, S., Kuchta, A., and Becla, E. (1952). Research on the nature of the noxious action of guinea-pig blood on the body louse. *Bull. Inst. mar. Med. Gdansk* 4, 104-7. (In Polish.)

Ledger, J. A. (1980) The arthropod parasites of vertebrates in Africa south of the Sahara. Volume IV. Phthiraptera (Insecta). *Publ. S. Afr. Inst. med. Res.* 56, pp. 1-327.

Lyal, C. H. C. (1985). Cladistics and classification of trichodectid mammal lice (Phthiraptera: Ischnocera). *Bull. Br. Mus. nat. Hist. (Ent.)* 51, 187-346.

—— (in prep. *a*). Fahrenholz's Rule and coevolution in insect-vertebrate associations, with particular reference to lice (Insecta: Phthiraptera).

—— (in prep. *b*). Host specificity in insect ectoparasites of vertebrates, with particular reference to lice (Phthiraptera).

—— (in prep. *c*). Coevolution of Trichodectidae (Insecta: Phthiraptera) and their mammalian hosts.

MacArthur, R. H. and Wilson, E. O. (1967). *The Theory of Island Biogeography*. Princeton University Press, Princeton, New Jersey.

Marshall, A. G. (1981). *The Ecology of Ectoparasitic Insects*. Academic Press, London.

Platnick N. I. (1977). Cladograms, phylogenetic trees, and hypothesis testing. *Syst. Zool.* 26, 438-42.

Plomley, N. J. B. and Thompson, G. B. (1937). Distribution of the biting louse. *Nature, Lond.* 140, 199.

Price, P. W. (1977). General concepts on the evolutionary biology of parasites. *Evolution, Lancaster, Pa.* 31, 405-20.

Price, R. D. and Beer, J. R. (1963). Species of *Colpocephalum* (Mallophaga: Menoponidae) parasitic upon the Falconiformes. *Can. Ent.* 95, 731-63.

Tarry, D. W. (1967). The occurrence of Linognathus setosus (Anoplura, Siphunculata) on poultry. *Vet. Rec.* 81, 641.

Thompson, G. B. (1940). The distribution of *Heterodoxus spiniger* (Enderlein). *Pap. Proc. R. Soc. Tasm.* 1939, 27-31.

Timm, R. M. (1983). Fahrenholz's rule and resource tracking: a study of host-parasite coevolution. In *Coevolution* (ed. M. H. Nitecki), pp. 225-65. University of Chicago Press, Chicago.

—— and Price, R. D. (1980). The taxonomy of *Geomydoecus* (Mallophaga: Trichodectidae) from the *Geomys bursarius* complex (Rodentia: Geomyidae). *J. Med. Entomol.* 17, 126-45.

Timmermann, G., von (1957). Stellung und Gleiderung der Regenpfeifervögel (Ordnung Charadriiformes) nach massgabe des Mallophagologischen befundes. In *First Symposium on Host Specificity among Parasites of Vertebrates, Neuchatel*, pp. 159–68. University of Neuchatel.

Traub, R. (1980). The zoogeography and evolution of some fleas, lice and mammals. In *Fleas. Proceedings of the International Conference on Fleas, Ashton Wold* (eds R. Traub and H. Starcke), pp. 93–172. A. A. Balkema, Rotterdam.

Webb, J. E. (1949). The evolution and host-relationships of the sucking lice of the Ferrungulata. *Proc. zool. Soc. Lond.* 119, 133–88.

Wiley, E. O. (1979). Cladograms and phylogenetic trees. *Syst. Zool.* 28, 88–92.

—— (1981). *Phylogenetics. The Theory and Practice of Phylogenetic Systematics*. Wiley-Interscience, John Wiley and Sons, New York.

Wilson, F. H. (1934). The life-cycle and bionomics of *Lipeurus heterographus* Nitzsch. *J. Parasit.* 5, 304–11.

6. Coevolutionary relationships of the helminth parasites of Australian marsupials

IAN BEVERIDGE

South Australian Department of Agriculture, Adelaide, South Australia

Abstract

The Australian marsupial fauna, which has evolved in relative isolation since the Tertiary period, provides a unique opportunity of studying coevolutionary relationships between the marsupials and their diverse helminth fauna. Examined from the point of view of the origins of elements of the helminth fauna, it appears likely that a few nematode and cestode groups may have been present in the original colonizing marsupials, while other parasite superfamilies or families clearly have been acquired from monotremes, birds, or reptiles present on the continent, and additional acquisitions have occurred from rodents or bats colonizing Australia from the north. At the generic of family level, evolution of the helminths has in many cases been parallel to that of the hosts, but there are also major shifts of helminths to distantly related groups. A similar pattern is evident when helminth evolution is examined at the species level, either by traditional morphological methods or by biochemical techniques. At all three levels therefore (the acquisition of helminth superfamilies, and evolution at both generic and species level) coevolution between host and parasite is mixed with much apparently random host-switching.

Introduction

The Australian continent has long been regarded by mammalogists as something of a laboratory of evolution, because of the way in which the marsupials, isolated since the Australian continent separated from Antarctica, have radiated to occupy most of the niches currently occupied by Eutherian mammals on other continents. While the phylogeny of the Australian marsupials has been for some time an interesting and stimulating field of scientific enquiry, the relationships of their prominent and diverse helminth parasite fauna have been, until recently, quite neglected. Of the 25 nematode superfamilies occurring in mammals (Anderson et al. 1974), 15 occur in Australian marsupials, and of the five cyclophyllidean cestode families occurring in terrestrial mammals, all are present in the marsupials. The trematode parasites of marsupials are less well known, but they are common in certain genera of the Dasyuroidea, with two genera of paramphistomes as well as the common liver fluke, *Fasciola hepatica*, occurring in the kangaroos. Acanthocephala are also known, but only from perameloid marsupials. The parasite fauna of marsupials is therefore extensive and extremely diverse. In spite of the large numbers of parasites known from marsupials, the origins of the parasite fauna, the degree to which it might or might not have coevolved with this extraordinary mammalian radiation and the ways in which such complex parasite communities developed have scarcely been examined. The literature on host–parasite relationships in the Marsupialia which does exist is scattered and fragmentary, and no previous attempt has been made to draw it together. This paper therefore attempts to summarize knowledge available to date on the distribution and evolution of the helminths of marsupials.

Hypotheses on the evolution of any parasitic group are only as sound as the taxonomic base upon which they are founded. In spite of the hundreds of helminth species already described from Australasian marsupials, a large fauna remains to be described. Because of this, any hypotheses must be presented with due caution and with the understanding that frequently there are unacceptable numbers of missing components to any evolutionary tree; components we can hope will be added in the years to come.

Origins of the marsupials

The most recent comprehensive account of the evolution and systematics of the Australian marsupials is that of Archer (1984). The Australian marsupial fauna is monophyletic (Kirsch 1977) and is probably derived from microbiotheriid marsupials, currently restricted to South America, which dispersed in the late Cretaceous or early Tertiary, colonizing Antarctica. The basic dasyuroid, perameloid, and diprotodont radiations of the Australian marsupials are thought to have diverged more than 45 million years ago, possibly before the Australian continent finally broke away from Antarctica. The Dasyuroidea is closest to the original South American invaders, consists exclusively of carnivores, and contains the families Dasyuridae (marsupial cats, Tasmanian devil, marsupial rats, and antechinuses), the Thylacinidae (with a single, possibly extinct species, the Tasmanian tiger) and the Myrmecobiidae (ant-eaters or numbats) (Fig. 6.1). The Perameloidea contains the omnivorous bandicoots ranged in two families, the Peramelidae, and Thalacomyidae.

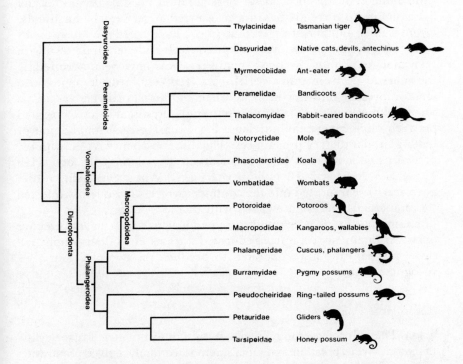

Fig. 6.1. Phylogenetic relationships of the families of Australian marsupials, modified from Archer (1984).

All of the herbivorous marsupials are grouped in the Diprotodonta, which is subdivided into three lineages, the Vombatoidea, containing the wombats (Vombatidae) and koala (Phascolarctidae), the Phalangeroidea containing the arboreal possums, phalangers, cuscus, gliders, and pygmy possums (Phalangeridae, Pseudocheiridae, Petauridae, Burramyidae), and the Macropodoidea containing the kangaroos and wallabies (Macropodidae, Potoroidae). The affinities of the marsupial mole, *Notoryctes typhlops*, are unclear, but Archer (1984) placed it between the Perameloidea and the Diprotodonta. The nomenclature of families of Australian marsupials used below follows that of Archer (1984) (Fig. 6.1). Coevolutionary relationships of the helminths of marsupials may conveniently be divided into three main sections, namely the origins of the parasite fauna, the broad coevolutionary patterns of host and parasite genera, and finally coevolution at the species level.

Origins of the parasite fauna of Australian marsupials

From a theoretical point of view, there are three possible origins for any helminth group currently parasitic in Australian marsupials. Firstly, the original microbiotheriid marsupials could have carried parasites with them from South America which subsequently radiated in parallel with their marsupial hosts. Secondly, parasites could have been acquired by the marsupials from other vertebrates (reptiles, birds, monotremes) already present on the continent when the marsupials arrived. This 'capture' mechanism (Chabaud 1965) would presumably have operated between phylogenetically unrelated hosts sharing the same habitat or having similar dietary preferences. The third possibility is that with the arrival from south-east Asia of bats about 20 million years ago (Hall 1984) and rodents 5–10 million years ago (Watts and Aslin 1981), parasites were imported into the continent which subsequently infected the marsupials and then evolved with them.

Examining the few helminth groups for which we can make either decisive statements or at least educated guesses as to their origins, all three mechanisms have operated to produce the complex fauna we recognize today.

The original immigrants

(a) The Herpetostrongylinae, a subfamily of trichostrongyloid nematodes widespread in marsupials, almost certainly arrived as parasites of the original invaders since parasite lineages with an obviously common ancestor exist in the present day South American didelphoid marsupials as well as in the Australian marsupials (Humphery-Smith 1983) (Fig. 6.2). In the Australian lineage, the most 'primitive' nematode species, that is with a synlophe or body ridge system consisting of three ventral ridges directed towards the left hand side, occur in the Dasyuroidea, the group most closely related to the South American didelphoid marsupials, and they have subsequently evolved to parasitize perameloids, phalangeroids, and macropodoids (Fig. 6.2). In the present day South American marsupials, representatives of the family Viannaiidae having a similar system of body ridges occur in the Didelphoidea, but have subsequently become parasites of certain South American rodents and undergone a major radiation within them (Durette-Desset 1971).

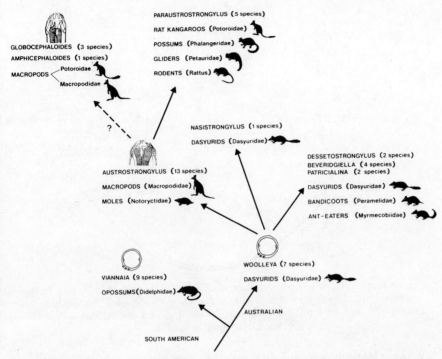

Fig. 6.2. Possible phylogenetic relationships of the Herpetostrongylidae (Nematoda: Trichostrongyloidea) in Australian marsupials, showing similarities in synlophe (body ridges) of the Viannaiidae of South American marsupials. Modified from Humphery-Smith (1981).

(b) The cestode subfamily Linstowiinae and in particular the subgenus *Linstowia* (*Paralinstowia*) is represented by species in bandicoots in Australia (*Linstowia semoni*) and in didelphoid marsupials in South America (*Linstowia jheringi*), suggesting again that this group of cestodes may have been brought to Australia by the invading marsupials (Fig. 6.3). However, the taxonomy and affiliations of this cestode genus are far from settled (Beveridge 1983) and any resultant zoogeographic hypotheses need to be treated with a degree of caution.

(c) Representatives of the trematode genera *Brachylaemus* and *Fibricola* occur in Australian dasyuroids and in South American didelphoids. Sandars (1957a,b) considered that in both cases the relevant trematodes were closely related, suggesting a common origin.

2. Acquisitions

A number of parasite groups appear to have been acquired by marsupials from vertebrates already present on the continent.

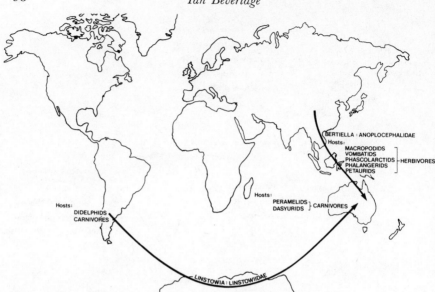

Fig. 6.3. Zoogeographic relationships of Australian anoplocephalid and linstowiid cestodes, indicating the possible origins of *Linstowia* (*Paralinstowia*) in South American marsupials, and a postulated south-east Asian origin for the *Bertiella — Progamotaenia* lineage of anoplocephalids.

(a) The trichostrongyloid family Mackerrastrongylidae has undergone a significant radiation within the monotreme family Tachyglossidae (Durette-Desset and Cassone 1983; Durette-Desset and Beveridge 1981), with the parasites of monotremes probably being derived ultimately from parasites of amphibians (Durette-Desset and Chabaud 1981). The subfamily Tasmanematinae in echidnas is considered as the origin of a major lineage of trichostrongyloid nematodes, the Mackerrastrongylinae, restricted to Dasyuroid and Perameloid marsupials (Durette-Desset and Chabaud 1981) (Fig. 6.4). In addition, one species of the genus *Nicollina* in monotremes has been acquired by the marsupial mole, *Notoryctes typhlops* (Beveridge and Durette-Desset, 1985).

(b) Another lineage of the Trichostrongyloidea, comprising the families Trichostrongylidae and Dromaeostrongylidae, were originally parasites of ratite birds (Durette-Desset and Chabaud 1981), a relatively ancient group with a Gondwanaland distribution (South America, Africa, New Zealand, Australasia). A genus in ostriches is thought to have given rise to the family Trichostrongylidae, which has subsequently evolved principally in herbivores of the Ethiopian region, while a related genus *Dromaeostrongylus* in Australian emus, appears to have infected perameloid and dasyuroid marsupials, and then subsequently evolved in parallel with the newly acquired host group (Durette-Desset and Chabaud 1981) (Fig. 6.5).

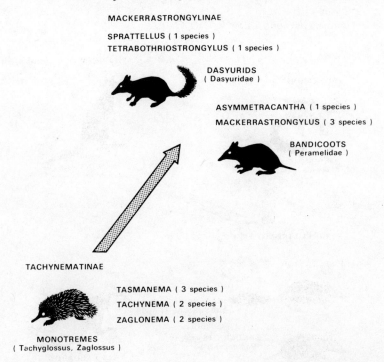

Fig. 6.4. Host and phylogenetic relationships of the Mackerrastrongylidae (Nematoda: Trichostrongyloidea) in Australasian monotremes and marsupials.

(c) Species of the linstowiid cestode genus *Oochoristica* occur in certain members of the dasyurid genera *Pseudantechinus* and *Dasykaluta*. The cestode genus also occurs widely in reptiles, and it has been suggested that the marsupials concerned may have acquired their parasites from reptiles (Beveridge 1977). *Oochoristica* is a cosmopolitan parasite of reptiles and occurs in South American marsupials as well as insectivores, rodents and primates. Because of this, any hypothesis on the origin of the Australian marsupial parasites must be guarded.

(d) *Linstowia* (*Linstowia*) occurs in monotremes in Australia (the echidna) with a single species in bandicoots (Peramelidae) (Beveridge 1983) and the latter may represent a transfer from monotremes to marsupials.

3. Recent arrivals

Several parasite groups have evidently reached Australia with rodents or bats invading the continent from south-east Asia. Most important numerically among these are the anoplocephaline cestodes and certain filarioid nematodes.

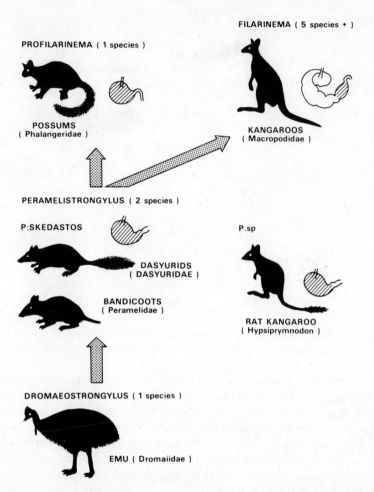

Fig. 6.5. Host and phylogenetic relationships of the Dromaeostrongylidae (Nematoda: Trichostrongyloidea) in Australian ratites and marsupials; localization within the stomach of the marsupial host is indicated by lined areas.

(a) The anoplocephaline cestode genus *Bertiella* occurs in south-east Asia in rodents, dermopterans and primates, in Africa in primates and rodents, and in South America in primates. In Australia, it occurs in rodents (*Uromys*, *Rattus*) and in arboreal folivorous marsupials belonging to the families Phalangeridae, Pseudocheiridae and Phascolarctidae (Beveridge 1976, 1982a, 1985). This zoogeographic distribution, plus the fact that among the species present in Australia those in rodents have many more plesiomorphic characters than those in marsupials (Beveridge

1985), suggests that the genus arrived with the rodents and subsequently invaded the marsupials (Figs 6.3 and 6.9). Beveridge (1982a) suggested that the invasion was dependent upon dietary preferences rather than host taxonomy, since apart from the (folivorous) Phalangeridae, only folivorous possums and gliders are parasitized and the koala which is also an arboreal folivore is only distantly related to the possums. However, the folivorous petaurids were subsequently placed in a distinct family, Pseudocheiridae, by Archer (1984) such that the distribution of the cestodes now follows more closely the taxonomy of the hosts. The remaining genera of anoplocephaline cestodes in marsupials (*Progamotaenia*, *Triplotaenia*) seem to be derived from *Bertiella*—like ancestors (Baer 1927; Beveridge 1976).

(b) The filarioid nematode genus *Breinlia* likewise occurs in rodents in south east Asia (Chabaud and Bain 1976), rodents in Australia (D. M. Spratt, pers. comm.) and various Australian marsupials (Spratt and Varughese 1975). *Breinlia* therefore may also have reached Australia from Asia with rodents, although an alternative route of entry from the south and subsequent migration from Australia to Asia has been suggested (Chabaud and Bain 1976; Bain et al. 1983).

(c) Two genera of spirurid nematodes, *Spirura* and *Echinonema* also have south east Asian affiliations, but in this case via bats. *Spirura aurangabadensis* occurs in bats and primates in south-east Asia, as well as in small dasyuroid marsupials in Australia (Spratt 1985) and is thought to have reached Australia either in bats or in the insect intermediate host. The genus *Echinonema*, in dasyuroid and perameloid marsupials in Australia, was considered, on the basis of larval morphogenesis, to be related to the genus *Seuratum* in bats, and bats were therefore proposed as the means by which the parasite genus reached Australia (Chabaud et al. 1980). As the intermediate host is an insect, a transfer from insectivorous bats as definitive hosts to insectivorous dasyurids and peramelids is highly plausible.

(d) The spirurid nematode genus *Synhimantus* occurs principally in birds. As a spirurid, an insect intermediate host is required, and it is therefore perhaps not surprising that the genus has established itself in certain instances in insectivorous mammals, with *S. australiensis* occurring in insectivorous dasyurid marsupials (*Antechinus*) (Beveridge and Barker 1975b) as well as in certain native rodents (Johnston and Mawson 1952). Another spirurid nematode, *Stammerinema suffodiax* which occurs in the stomachs of dasyurid marsupials (Beveridge and Barker 1975b) is derived phylogenetically, from the genus *Rusguniella* according to studies of the larval morphogenesis of *S. suffodiax* by Quentin and Beveridge (1986). *Rusguniella* species otherwise occur as parasites of birds, and *S. suffodiax* is considered a species derived by capture (Quentin and Beveridge 1986).

Of the various host groups that might have introduced parasites to the Australian continent, parasite introductions are clearly associated with rodents, bats and birds. The arrival of the dingo at least 4000 years ago (Gollan 1984), perhaps introduced by aboriginal man, and the more recent introduction of the feral cat and red fox by European man may perhaps have introduced the spirurid nematode genera *Cyathospirura* and *Cylicospirura*. Species of the first genus occur in dingoes, cats, and foxes as well as in the large dasyurid native cats (Clark 1981; Mawson 1968), but it is impossible at the moment to know whether these nematodes arrived with the carnivorous marsupials or later with the carnivorous placentals.

As intriguing as the successful parasitic introductions are, some of the failures are no less interesting. Although rodents may have played an important role in introducing certain groups of parasites to the continent, other parasites harboured by the rodents did not invade the marsupials. The Heligmosomatidae, a family of trichostrongyloid nematode occurring commonly in Australian rodents and which has its origins in south east Asia (Durette-Desset 1971) has made no incursion at all into the marsupials. By contrast some trichostrongyles of marsupials, notably *Woolleya hydromyos*, *Paraustrostrongylus ratti* and *Peramelistrongylus skedastos* have invaded rodents (Mawson 1961; Obendorf 1979). The only explanation which can be given at present is that the niches occupied by trichostrongyles were already filled in the marsupials by the endemic fauna. Such an explanation does not account, however, for the successful invasion of the rodents by certain marsupial trichostrongyles.

Not all parasites of marsupials fit neatly into the three schemes outlined above for their origins. *Baylisascaris tasmaniensis* is an ascaridoid nematode occurring in large dasyurid marsupials in Tasmania (Sprent 1970; Sprent *et al.* 1973). Congeners occur in Mustelidae, Procyonidae, and Ursidae principally in Holactic and Nearctic regions, none of which have any connections with marsupials. No simple hypothesis can plausibly explain this peculiar distribution at present (Sprent *et al.* 1973).

In summarizing the patterns of origin of the various helminth groups present in marsupials, it is evident that there is no correlation between the extent of a parasite radiation and the length of time it has been associated with a host group. The long period of association of the Herpetostrongylidae has produced a remarkable parasite radiation which seems lacking in the linstowiid cestodes. Furthermore, the anoplocephalid cestodes, which probably arrived relatively recently have undergone a substantial radiation, comparable with that of the Herpetostrongylidae.

The fact that so many helminth groups have apparently been acquired by the marsupials through 'host switching' is important in understanding subsequent coevolutionary pathways as similar mechanisms appear to have operated both at the generic and species level.

Coevolutionary patterns

Given the diverse origin of the parasite fauna, the question remains as to what extent and how these parasites have coevolved with their hosts. The parasites of the original invaders have sometimes undergone extensive periods of coevolution with their hosts, while in other parasite groups the degree of coevolution is less pronounced, and evolution of the parasites has apparently been determined by factors other than host relationships. Some examples serve to illustrate these points.

1. Herpetostrongylinae (Trichostrongyloidea)

The Herpetostrongylinae are one of the few trichostrongyloid groups which demonstrate, at the gross level, parallel evolution between hosts and parasites (Durette-Desset 1982). The phylogeny of this group was studied in detail by Humphery-Smith (1983) who recognized three distinct parasite lineages (Fig. 6.2). One, within the genus *Woolleya* which is restricted to Dasyuridae and Thylacinidae; a second lineage involving the genera *Dessetostrongylus*, *Beveridgiella*, and *Patricialina* which originated in Dasyuridae, and invaded Peramelidae and Myrmecobiidae; and a third lineage involving *Austrostrongylus* and *Paraustrostrongylus* which again originated as parasites of Dasyuridae, but managed to invade the herbivorous marsupials (Diprotodonta) and undergo a substantial development within them. This overall pattern of evolution closely follows that of the major host groups (Humphery-Smith 1983), however, the lineage within the diprotodont marsupials follows the coevolutionary plan less rigorously, and a tendency to misbehave is already evident among the parasites of Dasyuridae, with one unusual species, *Nasistrongylus antechini* quitting the intestine to take up residence in the nasal cavity of its dasyurid host (Beveridge and Barker 1975a). The Herpetostrongylinae entered diprotodont marsupials via the small intestine of the small wallabies belonging to the genus *Thylogale* (family Macropodidae) (Fig. 6.6). Some degree of development occurred in rock-wallabies (*Petrogale*), but the major development and radiation of *Austrostrongylus* has occurred in the wallabies belonging to *Macropus* (*Notomacropus*) and the related genus *Wallabia* (Fig. 6.6) (Beveridge and Durette-Desset 1986). Thus far, there is a close parallel between host and parasite evolution since the *Thylogale/Petrogale* group is older than *Macropus/Wallabia*, and because *Wallabia* itself may be ancestral to *Notomacropus* (Sanson 1978). However, the coevolutionary pattern subsequently breaks down. First of all, a species of *Austrostrongylus* has evolved in the marsupial mole, *Notoryctes typhlops*, a host probably unrelated to the diprotodonts (Strahan 1983), and this has been interpreted by Beveridge and Durette-Desset (1985) as a simple 'capture' in the sense of Chabaud (1965).

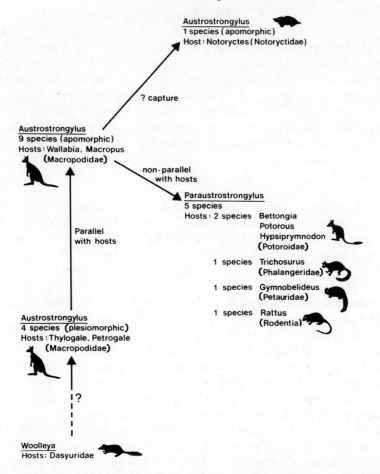

Fig. 6.6. Host and phylogenetic relationships of the nematode genera *Austrostrongylus* and *Paraustrongylus* (Trichostrongyloidea: Herpetostrongylidae).

Secondly, the genus *Paraustrostrongylus* is apparently derived from *Austrostrongylus* (Humphery-Smith 1981; Durette-Desset 1982), but instead of continuing to 'ascend' marsupial evolutionary lineages, it has taken the reverse course and 'descended'. *Par. hypsiprymnodontis*, *Par. bettongia* and *Par. potoroo* all occur in rat kangaroos of the Potoroidae, the older of the two families in the Macropodoidea. The hosts are small terrestrial mammals which browse on roots, fungi, tubers, and invertebrates (Calaby 1983). *Par. trichosuri* and *Par. gymnobelideus* occur in arboreal folivores and nectar feeders of the families Phalangeridae and Petauridae, respectively, in a lineage which predates the Macropodoidea. Finally, *Par. ratti* occurs in *Rattus fuscipes*, a completely unrelated host genus which has only been present on the Australian continent for about one million years (Watts and Aslin 1981). It appears therefore that this lineage, after invading the younger of the two macropodid families, underwent a period of parallel evolution with the hosts, yet with the emergence of the genus *Paraustrostrongylus* was suddenly freed of its host ties and invaded hosts irrespective of phylogenetic affiliation and, within the broad limits of herbivory, irrespective of diet.

The relationships of the Globocephaloidinae (*Globocephaloides*, *Amphicephaloides*) to the Herpetostrongylinae were not considered in detail by Durette-Desset and Chabaud (1981) or by Humphery-Smith (1983) largely because they lack a synlophe, the principal feature upon which the relationships within the family have been determined. Their specialized cephalic anatomy indicates a relationship with *Austrostrongylus* rather than with any other trichostrongyloid group and on this basis they are thought to be derived from *Austrostrongylus* (Fig. 6.2). If this is the case, then their evolution has taken place entirely within the Macropodidae.

2. Dromaeostrongylidae

The evolution of the trichostrongyloid family Dromaeostrongylidae is of particular interest from the point of view of its localization within the host. The family apparently derives from parasites of the small intestine of ratite birds (Durette-Desset and Chabaud 1981), but is first encountered in marsupials in the genus *Peramelistrongylus* occurring in the stomachs of Dasyuridae and Peramelidae (Fig. 6.5). The genus *Profilarinema* clearly has a common ancestry with *Peramelistrongylus*, but occurs in the stomach of the brush tailed possum, *Trichosurus vulpecula* (Phalangeridae).

Within the Macropodoidea, one member (*Hypsiprymnodon*) is monogastric and harbours a species of *Peramelistrongylus*, while in the remainder of the macropodoids, the stomach has evolved into a large, complex, sacculated organ, reminiscent in certain respects of the fore-stomachs of ruminants (Hume 1978). Concomitant with this development, the acid-secreting region has been reduced to a small component adjacent to the pylorus. Inhabiting this small niche in the Macropodidae is the nematode genus *Filarinema* clearly derived from a common ancestor with *Profilarinema* and occupying a comparable niche, in spite of the enormous changes in the morphology of the stomach that have occurred in the evolution of the hosts.

This lineage represents not only an example of parallel evolution with the hosts, but also the importance of the site of localization of a parasite within a host.

3. Strongyloidea

For the Strongyloidea of marsupials, the evolutionary changes occurring in the digestive tracts of the hosts have been a major factor in parasite evolution. The Strongyloidea is a cosmopolitan nematode superfamily occurring mainly in sites of fermentative digestion (essentially the large intestines) of ratite birds, tortoises, perissodactyls, artiodactyls, rodents, and primates. The group appears in the Diprotodonta without any ancestors in the carnivorous marsupials, and one can only assume that the marsupials have acquired these particular nematodes from another group of Australian vertebrates. Ratite birds have been suggested as a possible origin (Beveridge 1982b) since ratites are common in Australia and occur as hosts of the group on other continents, however, no strongyloids have been found so far in Australian ratites. The most primitive of the Australian genera occur in the large intestine of Macropodoidea, with one lineage subsequently invading wombats, occurring in the colon, the major site of fermentative digestion in the tract of wombats (Fig. 6.7). In the Macropodidae, with the development of the sacculated stomach, new niches were opened up for strongyloid nematodes. *Hypsiprymnodon moschatus* is the only monogastric species known and possesses a simple stomach akin to that found in members of the Phalangeroidea (Pearson 1950). The only strongyloid known from *Hypsiprymnodon*, *Corollostrongylus hypsiprymnodontis*, occurs in the caecum, colon and ileum of its host (Beveridge 1978) (Fig. 6.7). The potoroid stomach is large, complex, and sacculated with a relatively enormous anterior blind sac, while the macropodid stomach has a small blind sac and an elongate, tubular, haustrated mid-stomach (Langer 1979a, 1980; Langer et al. 1980). Fermentative digestion occurs in the macropodid stomach, but to a much lesser degree in the potoroid stomach (Langer 1979b), and this is correlated to some extent with diet, since potoroids feed principally on roots, succulent vegetation, fungi, and grasses, while macropodids are grazers or browsers (Langer 1980). Probably the first strongyloid nematodes to take advantage of this new site of fermentative digestion in the sacculated macropodoid stomach were ancestors of the related genera *Macropostrongyloides* and *Paramacropostrongylus* which, apart from being intermediate morphologically between the 'primitive' genera inhabiting the caecum, and the stomach-inhabiting Cloacininae, occur either in the caeca or in the stomachs of their hosts (Fig. 6.7). They may have had ancestors in common with the sub-family Cloacininae which is restricted to the stomach and oesophagus (Fig. 6.7) and has undergone an explosive radiation within the macropodid stomach producing some 33 genera and over 100 species. Characteristic of infections in large macropodids are vast parasite burdens [up to 300 000 nematodes (Beveridge and Arundel 1979)] and the presence

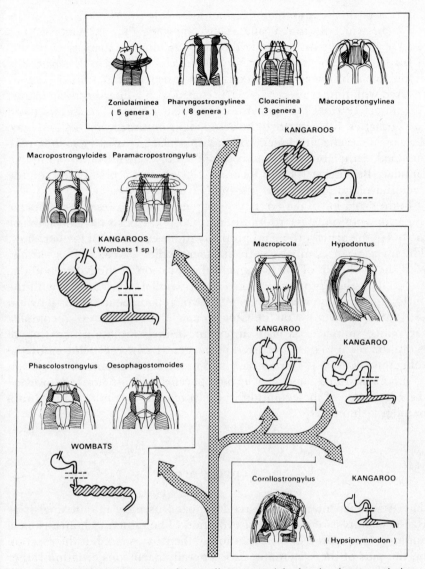

Fig. 6.7. Strongyloid nematodes of Australian marsupials showing host associations, localization in the host (shaded areas), and possible phylogenetic relationships.

of species flocks, particularly in the genus *Cloacina* (see Beveridge 1979c).

In contrast to the parasitic explosion in the macropodids, the potoroids were apparently by-passed, with only a single genus of two species occurring in them. Furthermore, this genus has apomorphic characters which suggest that it has invaded potoroids from macropodids (Beveridge 1982b). The lack of strongyloid nematodes in potoroids may be related to differences in the amount of fermentative digestion occurring in the stomach, which is much smaller in potoroids than in macropodids (Langer 1980) or may be due to as yet unknown causes.

In the kangaroo subgenus *Notomacropus* as well as in *Wallabia*, the oesophageal lining is not smooth as in most other mammals, but is covered with numerous tiny papillae coated with a thick bacterial plaque (Obendorf 1984a). Two genera of the Cloacininae, *Cyclostrongylus* and *Spirostrongylus*, have invaded the oesophagus of these wallabies and have developed specific attachment mechanisms (alae) to assist them to remain attached, either between rows of papillae or coiled around clumps of papillae (Beveridge 1982c; Obendorf 1984b), and probably feed on the bacterial plaque (Obendorf 1984b).

In the evolution of the Strongyloidea therefore, the sites of fermentative digestion within the gastro-intestinal tracts of the hosts have been major factors in the distribution and probably in the evolution of the parasites. The invasion of Vombatidae from Macropodidae is clearly not parallel with the evolution of the hosts, and the invasion of the Macropodidae rather than the Potoroidae is also not a parallel development with the hosts, since the Potoroidae predate the Macropodidae in the fossil record (Stirton *et al.* 1968). For the Cloacininae, the evolution of a complex sacculated stomach has been important in their radiation; 'centres' of evolution seem to have occurred in the genus *Thylogale* followed by the subgenus *Macropus* (*Notomacropus*), and in this limited sense the evolution of the Cloacininae has been in part parallel with its hosts. However, the remainder of the evolution in the subfamily has not occurred in a parallel fashion.

Convergence

The Australian marsupials have long been used by mammalogists as classic examples of convergent evolution. The same mechanisms have apparently operated in the evolution of their parasites, but the reaction on the part of the systematist has generally been consternation rather than admiration. Because helminths are relatively simple organisms with a limited repertoire of taxonomic characters, convergence may be somewhat harder to detect than in vertebrates, and taxonomic problems have arisen from the failure to detect it.

(1) *Hypodontus macropi* was initially described as a new 'hookworm' from the large intestine of the red kangaroo. It had cutting plates around the mouth opening, with a deviated head typical of the Necatorinae (hookworms) of eutherian mammals, and remained within the Ancylostomatoidea (hookworms) for 40 years following its description.

Inglis (1968), however, pointed out that the head was deviated ventrally and not dorsally as in the remainder of the Ancylostomatoidea and Beveridge (1979a) listed a suite of characters which aligned it with the Strongyloidea rather than the Ancylostomatoidea. Lichtenfels (1980) removed the genus from the Ancylostomatoidea to the Strongyloidea, and it was placed near the base of the Strongyloid lineage in Australian marsupials by Beveridge (1982b). *Hypodontus* is therefore an excellent example of convergent evolution in nematodes.

(2) *Globocephaloides* is a genus occurring in the small intestine of macropodids which, because of its globular buccal capsule and prominent teeth bears a superficial resemblance to the hookworm genus *Globocephalus* of pigs. *Globocephaloides* was initially placed in the Ancylostomatoidea, and the genus at one stage was even synonymized with *Globocephalus*, until Inglis (1968) showed that *Globocephaloides* had evolved from other trichostrongyloid parasites of marsupials in convergence with *Globocephalus*. Inglis' views were substantiated subsequently by other workers (Beveridge 1979b; Durette-Desset and Chabaud 1981).

(3) Less striking examples of the development of convergent characteristics occur in the strongyloid genera *Cyclostrongylus* and *Spirostrongylus* which have developed alae to assist in their attachment to the oesophageal lining of certain macropodids. Comparable developments are widespread in the Trichostrongyloidea, but these two genera are the only strongyloids to have developed such an attachment mechanism.

(4) *Woodwardostrongylus* is a genus of strongyloid nematode which lives in tunnels in the gastric and oesophageal mucosa of macropodids. While this mode of existence is common in capillariids and in certain spirurid parasites of birds, *Woodwardostrongylus* is the only strongyloid genus known to occur in this site in its host.

(5) Convergence in cestodes may be particularly difficult to detect. *Anoplotaenia dasyuri* is a common parasite of the Tasmanian devil (*Sarcophilus harrisii*) and was placed in the cestode family Taeniidae on the basis of the morphology of the mature proglottis and the uterus. The Taeniidae, which is thought to be a family of recent origins (Rausch 1981), is essentially parasitic in Carnivora, so the presence of a taeniid in a marsupial is not easily explained in a zoogeographical sense. Studies on the life cycle of the parasite revealed initially that the use of a mammal (wallabies) as intermediate host as well as the ultrastructure of the egg envelopes and the hatching mechanism of the egg were all consistent with its position in the Taeniidae (Beveridge *et al.* 1975). However, the morphogenesis of the metacestode was clearly non-taeniid and followed the pattern already known for the Linstowiidae, a family occurring in reptiles and mammals, but with a metacestode in arthropods. Metacestode morphogenesis is an extremely important taxonomic

criterion (if it is known) (Freeman 1973), and in this case it aligns *Anoplotaenia* with the Linstowiidae, which occur in other dasyurid and in peramelid marsupials (Beveridge 1982*a*), rather than with the Taeniidae. *Anoplotaenia* may therefore represent a remarkable example of convergent evolution in cestodes, and may indicate a major evolutionary radiation within the Linstowiidae of marsupials the details of which we are yet to discover.

Coevolutionary mechanisms at the species level

Examples of host-parasite coevolution given above are at the generic level or above, both for parasite and for host. A fairly gross picture of coevolution is therefore being presented, and the frequent individual exceptions or inconsistencies that one has to point out or ignore when making generalizations raise the question as to how broad coevolutionary patterns operate at the species level. Are the parasite radiations occurring, species by species, exactly in parallel with host evolution, or is their evolution more complex and less ordered? In order to examine this problem, one requires parasites with a fairly high degree of host specificity, and these are relatively few among the parasite of marsupials. However, the example selected below may indicate a general pattern and probably reflects as much information as one can obtain from simple morphological studies.

Species of the cestode genus *Bertiella* are highly host specific (Beveridge 1976, 1985), each cestode species occurring in one species of host (except *B. trichosuri*). A cladogram representing the relationships of the parasites, derived from Beveridge (1985) is shown in Fig. 6.8. Four of the major clades are each restricted to a particular marsupial host family (Beveridge 1985), one clade (A) occurs in rodents, and one (clade C) is restricted to a particular host subgenus, *Pseudocheirus* (*Pseudochirops*). In addition, in several species pairs, their is a close relationship between their relative hosts. *B. campanulata* and *B. paucitesticulata* (Fig. 6.8) both occur in hosts belonging to *Pseudocheirus* (*Pseudochirops*). Similarly, *B. mawsonae* and *B. vesicularis* (Fig. 6.8) occur, respectively, in *Petauroides* and *Hemibelideus*, two taxa which are considered to be monophyletic (Archer 1984).

The cladogram therefore demonstrates that parallel evolution has occurred between some hosts and parasites, but it also shows evidence of host switching by parasites. For example, *B. foederata* which occurs in *Pseudocheirus peregrinus* (clade B) belongs to a completely different clade to the other parasites of this and related hosts (*B. affinis*, *B. aberrata*, *B. paraberrata* in clade E), resembling more closely *B. petaurina*, a parasite of *Petauroides volans*. The same clade (B) contains parasites of Phalangeridae, Pseudocheiridae, and even the koala (Phascolarctidae),

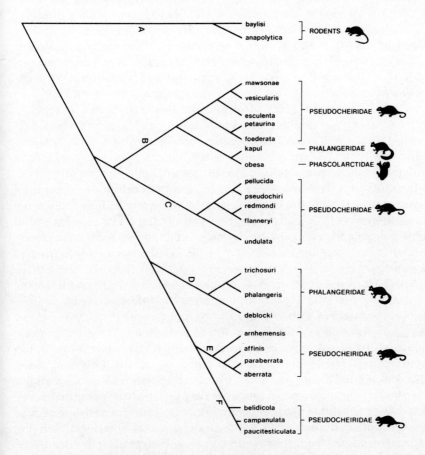

Fig. 6.8. Host and phylogenetic relationships of Australasian species of the anoplocephalid cestode genus *Bertiella*. Modified from Beveridge (1985).

a distribution which can only be explained by host switching. In so far as the evolution of the genus *Bertiella* can be traced on morphological grounds, therefore, there is a mosaic of parallel evolution with the host and of host switching.

The taxonomy of the bile duct cestodes of marsupials (*Progamotaenia* spp.) has been investigated recently using iso-enzyme analyses and applying the techniques to the parasites and their hosts at the same time. *Progamotaenia festiva* appears on the available morphological evidence to be a polymorphic species, with slightly different morphological 'forms' occurring in different macropodid and vombatid host species (Beveridge 1976). However, where different host species are sympatric, there is apparently no interchange of the morphological 'forms' of this cestode, suggesting a degree of host specificity. Analysis of parasite iso-enzymes (Baverstock *et al.* 1985) has revealed that the different morphological 'forms' described by Beveridge (1976) are indeed genetically distinct and are often host specific. The study revealed in addition a variety of genetically distinct groups that had not been recognized morphologically. Most were host specific, some over large geographic areas, but at least three clear examples of host switching were detected. In *Macropus robustus* from north Queensland, two genetically distinct forms were present, frequently in the same host individual, and the fact that the genetic differences were maintained even between specimens lying side by side in the same bile duct was taken as strong evidence that the genetically different populations being studied were in fact distinct biological species. When the relationships of the cestodes were compared with those established by the same techniques for the hosts, there was little correlation between the two. The sole exception occurred in the wombats, where the two cestode species were more closely related to one another than to any of the cestode species occurring in kangaroos.

Apart from the fundamental problems raised by this study in the recognition of sibling species in cestodes, the information it provides on host–parasite evolution is similar to that provided by more simple morphological studies, namely that within any given parasite taxon, there is a mixture of sublineages in some of which parasite evolution has occurred in parallel with the hosts and in some of which host switching between related hosts or even unrelated hosts has been a major evolutionary mechanism. The mixture between the two is not predictable for any lineage and is not homogeneous in its occurrence within a single lineage. More studies need to be undertaken at this level, but the impression gained from the marsupial parasites thus far is that the broad coevolutionary relationships seen at the gross level conceal a much less coherent pattern of coevolution at the species level. This is evident in the major lineages only where large host switches occur (e.g. Dromaeostrongylidae), or where completely 'aberrant' species of parasite have evolved within a taxon (e.g. *Nasistrongylus*).

Conclusion

The relative simplicity of the Australian marsupial fauna in being monophyletic in origin (Kirsch 1977) and having radiated in isolation for nearly 50 million years is not therefore reflected in its complex parasite fauna which has a number of origins, including parasites of South American marsupials, transfers from other host groups (monotremes, ratite birds, reptiles) during a long period of relative isolation and introductions by bats, rats, and birds from south-east Asia. Within their marsupial hosts, the evolution of the parasites has been a patchwork of varying degrees of coevolution interspersed with major shifts of host group. Examples of close coevolution in the marsupial parasite fauna certainly exist, but equally striking are the radiations that have occurred across largely unrelated host groups, and in addition, just as convergent evolution of the hosts has, in the past, confused mammal taxonomists, so convergent evolution in the parasites has contributed to the difficulties of treating their systematics. The complex evolutionary history of the marsupial helminths contrasts sharply with studies such as those undertaken on the oxyurids of primates (Cameron 1929; Brooks and Glen 1982) in which highly specific parasites have evolved precisely in parallel with their hosts and even allow speculation about host relationships from those of the parasites. The data derived from marsupials would seem to offer no such avenues and indicate that generalizations about parasite evolution based on studies of a single group are extremely hazardous. Evolutionary pathways are rarely duplicated in the helminth communities of marsupials and do not seem to conform readily to the classical rules of parasite evolution.

Finally, a cautionary note is needed. The parasite fauna of marsupials remains relatively little studied, with many undescribed taxa awaiting description, and the analysis of evolutionary lineages has scarcely begun. The results presented here have to be viewed as preliminary, but suggest that it will prove to be a field of singular interest in the future.

Acknowledgements

Thanks are due to Dr D. M. Spratt for reading a draft of the manuscript and for allowing use of his unpublished data, to Mr M. S. Hullam for preparing the illustrations, and to Dr M. Archer for permitting the use of a modified version of one of his published figures indicating marsupial relationships.

References

Anderson, R. C., Chabaud, A. G., and Willmott, S. (1974). *CIH Keys to the Nematode Parasites of Vertebrates*. No. 1. *General Introduction, Glossary of Terms and Keys to Subclasses, Orders and Superfamilies*. Commonwealth Agricultural Bureaux, Farnham Royal, England, pp. 17.

Archer, M. (1984). The Australian marsupial radiation. In *Vertebrate Zoogeography and Evolution in Australasia* (eds M. Archer and G. Clayton), pp. 633–808, Hesperian Press, Western Australia.

Bain, O., Baker, M., and Chabaud, A. G. (1983). Nouvelles données sur la lignée *Dipetalonema*. *Ann. Parasit. hum. comp.* 57, 593-620.

Baer, J. G. (1927). Monographie des Cestodes de la famille des Anoplocephalidae. *Bull. biol. Fr. Belg.* Suppl. 10.

Baverstock, P. R., Adams, M., and Beveridge, I. (1985). Biochemical differentiation in bile duct cestodes and their marsupial hosts. *Mol. Biol. Evol.* 2, 321-37.

Beveridge, I. (1976). A taxonomic revision of the Anoplocephalidae (Cestoda: Cyclophyllidea) of Australian marsupials. *Aust. J. Zool., Suppl. Ser.* No. 44, 1-110.

—— (1977). On two new Linstowiid cestodes from Australian marsupials. *J. Helminthol.* 51, 3-40.

—— (1978). *Corollostrongylus hypsiprymnodontis* gen. et. sp. n. (Nematoda: Strongylidae) from the rat kangaroo, *Hypsiprymnodon moschatus* (Marsupialia). *J. Parasitol.* 64, 657-60.

—— (1979a). *Hypodontus macropi* Moennig, 1929, a hookworm-like parasite of macropodid marsupials. *J. Helminthol.* 53, 229-44.

—— (1979b). A review of the Globocephaloidinae Inglis (Nematoda: Amidostomatidae) from macropodid marsupials. *Aust. J. Zool.* 27, 151-75.

—— (1979c). Species of *Cloacina* Linstow, 1898 (Nematoda: Strongyloidea) from the black-tailed wallaby, *Wallabia bicolor* (Desmarest, 1804) from eastern Australia. *J. Helminthol.* 53, 363-78.

—— (1982a). Specificity and evolution of the anoplocephalate cestodes of marsupials. *Mém. Mus. natn. Hist. nat., Paris, sér. A., Zool.* 123, 103-9.

—— (1982b). Evolution of the strongyloid nematodes of Australian marsupials. *Mém. Mus. natn. Hist. nat., Paris, sér. A, Zool.* 123, 87-92.

—— (1982c). A taxonomic revision of the Pharyngostrongylinea Popova (Nematoda: Strongyloidea) from macropodid marsupials. *Aust. J. Zool. Suppl. Ser.* No. 83: 1-150.

—— (1983). The genus *Linstowia* Zschokke, 1899 (Cestoda: Anoplocephalidae) in Australian mammals with the description of a new species, *L. macrouri*. *Syst. Parasitol.* 5, 291-304.

—— (1985). The genus *Bertiella* (Cestoda: Anoplocephalidae) from Australasian mammals: new species, new records and redescriptions. *Syst. Parasitol.* 7, 241-89.

—— and Arundel, J. H. (1979). Helminth parasites of grey kangaroos, *Macropus giganteus* Shaw and *M. fuliginosus* (Desmarest), in eastern Australia. *Aust. Wildl. Res.* 6, 69-77.

—— and Barker, I. K. (1975a). *Nicollina antechini* sp. nov. (Nematoda: Amidostomatidae) from the nasal cavity of the dasyurid marsupial *Antechinus stuartii* Macleay, 1841, and associated pathology. *J. Parasitol.* 6, 489-93.

—— and —— (1975b). Acuariid, capillariid and hymenolepidid parasites of the dasyurid marsupial *Antechinus stuartii* Macleay, 1841, from south eastern Australia. *J. Helminthol.* 49, 211-27.

—— and Durette-Desset, M.-C. (1985). Two new species of nematode (Trichostrongyloidea) from the marsupial mole, *Notoryctes typhlops* (Stirling). *Bull. Mus. natn. Hist. nat., Paris, 4e sér.* 7, 341-7.

—— and —— (1986). New species of *Austrostrongylus* Chandler, 1924 (Nematoda: Trichostrongyloidea), from Australian marsupials with a redescription of *A. minutus* Johnston and Mawson, 1938 and description of a new genus, *Sutarostrongylus*. *Bull. Mus. natn. Hist. nat.*, Paris, 4e sér. (in press).

—— Rickard, M. D., Gregory, G. G., and Munday, B. L. (1975). Studies on *Anoplotaenia dasyuri* Beddard, 1911 (Cestoda: Taeniidae), a parasite of the Tasmanian devil: observations on the egg and metacestode. *Int. J. Parasitol.* 5, 257–67.

Brooks, D. R. and Glen, D. R. (1982). Pinworms and primates: a case study in coevolution. *Proc. Helm. Soc. Wash.* 49, 76–85.

Calaby, J. C. (1983). Potoroos, Bettongs and Rat-kangaroos, Family Potoroidae. In *Complete Book of Australian Mammals* (ed. R. Strahan), pp. 177–8. Angus and Robertson, London.

Cameron, T. W. M. (1929). The species of *Enterobius* Leach, in primates. *J. Helminthol.* 7, 161–82.

Chabaud, A. G. (1965). Spécificité parasitaire. In *Traité de Zoologie*. Vol 4. *Classe des Nématodes* (ed. P. P. Grassé) pp. 548–57, Masson et Cie, Paris.

—— and Bain, O. (1976) La lignée *Dipetalonema*. Nouvel essai de classification. *Ann. Parasit. hum. comp.* 51, 365–97.

—— Serreau, C., Beveridge, I., Bain, O., and Durette-Desset, M.-C. (1980). Sur les nématodes Echinonematinae. *Ann. Parasit. hum. comp.* 55, 427–43.

Clark, W. C. (1981). *Cylicospirura advena* n. sp. (Nematoda: Spirocercidae) a stomach parasite from a cat in New Zealand, with observations on related species. *Syst. Parasitol.* 3, 185–91.

Durette-Desset, M.-C. (1971). Essai de classification des nématodes Héligmosomes. Corrélation avec la paléobiogéographie des hôtes. *Mém. Mus. natn. Hist. nat.*, Paris, sér. A, Zool. 49, 1–126.

—— (1982). Relations hôtes-parasites chez les Trichostrongyloides *Mém. Mus. natn. Hist. nat.*, Paris, sér. A, Zool. 123, 93–100.

—— and Beveridge, I. (1981). *Zaglonema* n. gen. (Nematoda: Trichostrongyloidea) parasite de *Zaglossus bruynii* de Nouvelle-Guinée. *Ann. Parasit. hum. comp.* 56, 67–72.

—— and Cassone, J. (1983). A taxonomic revision of the trichostrongyloid nematode parasites of the echidna, *Tachyglossus aculeatus* (Monotremata). *Aust. J. Zool.* 31, 257–84.

—— and Chabaud, A. G. (1981). Nouvel essai de classification des nématodes trichostrongyloides. *Ann. Parasit. hum. comp.* 56, 297–312.

Freeman, R. S. (1973). Ontogeny of cestodes and its bearing on their phylogeny and systematics. *Adv. Parasitol.* 11, 481–557.

Gollan, K. (1984). The Australian Dingo: in the shadow of man. In *Vertebrate Zoogeography and Evolution in Australasia* (eds M. Archer and G. Clayton), pp. 921–7. Hesperian Press, Western Australia.

Hall, L. (1984). And then there were Bats. In *Vertebrate Zoogeography and Evolution in Australasia* (eds M. Archer and G. Clayton), pp. 838–52. Hesperian Press, Western Australia.

Hume, I. D. (1978). Evolution of the Macropodidae digestive system. *Aust. Mammal.* 2, 37–42.

Humphery-Smith, I. (1981). *Paraustrostrongylus gymnobelideus* n. sp. (Nematoda: Trichostrongyloidea) parasite de *Gymnobelideus leadbeateri* (Marsupialia, Petauridae) de Victoria, Australie. *Bull. Mus. natn. Hist. nat., Paris, 4ᵉ sér.* 3, 509–13.
—— (1983). An hypothesis on the evolution of Herpetostrongylinae (Trichostrongyloidea: Nematoda) in Australian marsupials, and their relationships with Viannaiidae, parasites of South American marsupials. *Aust. J. Zool.* 31, 931–42.
Inglis, W. G. (1968). The geographical and evolutionary relationships of Australian trichostrongyloid parasites and their hosts. *J. Linn. Soc. (Zool.)* 47, 327–47.
Johnston, T. H. and Mawson, P. M. (1952). Some nematodes from Australian birds and mammals. *Trans. R. Soc. S. Aust.* 75, 30–7.
Kirsch, J. A. W. (1977). The comparative serology of the Marsupialia, and a classification of Marsupials. *Aust. J. Zool. Suppl. Ser.* No. 52, 1–152.
Langer, P. (1979a). Phylogenetic adaptation of the stomach of the Macropodidae Owen, 1839, to food. *Zeit. Saugetierkd.* 44, 321–33.
—— (1979b). Functional anatomy and ontogenetic development of the stomach in the macropodine species *Thylogale stigmatica* and *Thylogale thetis* (Mammalia: Marsupialia). *Zoomorphologie* 93, 137–51.
—— (1980). Anatomy of the stomach in three species of Potoroinae (Marsupialia: Macropodidae). *Aust. J. Zool.* 28, 19–31.
—— Dellow, D. W., and Hume, I. D. (1980). Stomach structure and function in three species of macropodine marsupials. *Aust. J. Zool.* 28, 1–18.
Lichtenfels, J. R. (1980). *CIH Keys to the Nematode Parasites of Vertebrates.* No. 7. *Keys to Genera of the Superfamily Strongyloidea.* Commonwealth Agricultural Bureaux, Farnham Royal, England.
Mawson, P. M. (1961). Trichostrongyles from rodents in Queensland, with comments on the genus *Longistriata* (Nematoda: Heligmosomatidae). *Aust. J. Zool.* 9, 791–826.
—— (1968). Two new species of Nematoda (Spirurida: Spiruridae) from Australian dasyurids. *Parasitology* 58, 75–8.
Obendorf, D. L. (1979). The helminth parasites of *Rattus fuscipes* (Waterhouse) from Victoria, including description of two new nematode species. *Aust. J. Zool.* 27, 867–79.
—— (1984a). The macropodid oesophagus. I. Gross anatomical, light microscopic, scanning and transmission electron microscopic observations of its mucosa. *Aust. J. Zool.* 32, 415–36.
—— (1984b). The macropodid oesophagus. III. Observations on the nematode parasites. *Aust. J. Zool.* 32, 437–46.
Pearson, J. (1950). The relationship of the Potoroidae to the Macropodidae (Marsupialia). *Pap. Proc. R. Soc. Tas.* 1949, 212–29.
Quentin, J.-C. and Beveridge, I. (1986). Comparative morphogenesis of the cephalic structures of the acuariid nematodes *Stammerinema soricis* (Tiner, 1951), *Antechiniella suffodiax* (Beveridge and Barker, 1975) comb. et. gen. nov., and *Skrjabinoclava thapari* (Teixeira de Freitas, 1953). *Syst. Parasitol.* (in press).
Rausch, R. L. (1981). Morphological and biological characteristics of *Taenia rileyi* Loewen, 1929 (Cestoda: Taeniidae). *Can. J. Zool.* 59, 653–66.

Sandars, D. F. (1957a). On *Brachylaemus* (Trematoda) from Marsupials. *J. Helminthol.* 31, 265–72.
—— (1957b). A new strigeid trematode from an Australian marsupial. *J. Helminthol.* 31, 257–64.
Sanson, G. D. (1978). The evolution and significance of mastication in the Macropodidae. *Aust. Mammal.* 2, 23–8.
Spratt, D. M. (1985). *Spirura aurangabadensis* (Ali and Lovekar) (Nematoda: Spiruridae) from small Dasyuridae (Marsupialia). *Trans. R. Soc. S. Aust.* 109, 25–29.
—— and Varughese, G. (1975). A taxonomic revision of filarioid nematodes from Australian marsupials. *Aust. J. Zool. Suppl. Ser.* No. 35, 1–99.
Sprent, J. F. A. (1970). *Baylisascaris tasmaniensis* sp. nov. in marsupial carnivores: heirloom or souvenir? *Parasitology* 61, 75–86.
—— Lamina, J., and McKeown, A. (1973). Observations on migratory behaviour and development of *Baylisascaris tasmaniensis. Parasitology* 67, 67–83.
Stirton, R. A., Tedford, R. A., and Woodbourne, M. O. (1968). Australian Tertiary deposits containing terrestrial mammals. *Univ. Calif. Publ. Geol. Sci.* 77, 1–30.
Strahan, R. (1983). The Australian Mammal Fauna. In *Complete Book of Australian Mammals* (ed. R. Strahan), pp. xvii–xxi. Angus and Robertson, London.
Watts, C. H. S. and Aslin, H. J. (1981). *The Rodents of Australia.* Angus and Robertson, London.

7. Patterns in coevolution

JOHN N. THOMPSON

Department of Botany and Zoology,
Washington State University, U.S.A.

Abstract

In this paper five patterns in coevolution that integrate population biology, systematics, and geographic variation in interactions are considered.

(1) Contrary to recent suggestions, the current data on Umbelliferae and insects do not suggest that evolutionary changes in coumarin chemistry have produced a pattern of coevolution that follows closely the Ehrlich–Raven scenario.

(2) Cospeciation, that is reciprocal speciation caused by interactions between species, may be limited to very specific and intimate kinds of interactions, such as those between agaonid wasps and figs.

(3) *Mixed-process coevolution* may be defined as reciprocal evolution of interacting species in which adaptation of a population of one species to a population of a second species causes the population of the second species to become reproductively isolated from other populations. Mixed-process coevolution may be most common in interactions in which (a) one species directly transmits the gametes of another species or (b) a maternally-inherited mutualist or commensal of one host population is unnecessary or even virulent in other populations.

(4) The arms race analogy for coevolution of antagonists is an overly restrictive view and is only one of at least five possible outcomes of coadaptation in antagonists.

(5) The outcome of coevolution is likely to depend on age, size, social, and demic structure of populations, but these effects have been incorporated only sparingly into current coevolutionary theory.

©The Systematics Association, 1986. This chapter is from *Coevolution and systematics* (eds A. R. Stone and D. L. Hawksworth) published for the Systematics Association by the Clarendon Press, Oxford.

Finally, it is argued that analyses of patterns in coevolutionary change will demand studies of *interaction norms* which are variations in outcome of interactions across different environments as genotypes are held constant.

Introduction

During the past 20 years a rich variety of hypotheses has been suggested on patterns in coevolution: on the ecological conditions likely to favour coevolution, on the genetics of coevolution, and on patterns in the direction of coevolutionary change. These hypotheses, some derived from empirical studies and others from general theory in population and evolutionary biology, allow us a developing framework within which to study coevolution and test expectations.

Here, some current questions, hypotheses, and approaches on patterns in coevolution that require the integration of studies in systematics, population biology, and geographical variation will be discussed. We are slowly developing a broad framework on how interactions evolve, and when and where coevolution is likely, but the framework will still require us to sort among multiple hypotheses as we continue to search for patterns. I do not intend to be exhaustive in my list of current questions, hypotheses, and approaches. Elsewhere I have considered how different modes of interaction—parasitism, grazing, predation, competition, symbiotic mutualism, non-symbiotic mutualism—differ in the likelihood and pattern of coevolution (Thompson 1982). Here five related problems in coevolution are considered: (1) the need to incorporate systematics, biogeography of interactions, and population biology in evaluating the Ehrlich–Raven formulation of coevolution; (2) the relationship of coevolution and speciation; (3) the limited utility of the arms race analogy in studying patterns in coevolution; (4) the relationship of coevolution and population structure; and (5) the need to understand reaction norms of interactions. I also introduce two new terms, 'mixed-process coevolution' and 'interaction norms', that may be useful in focusing coevolutionary studies.

Coevolution is reciprocal evolutionary change in interacting species. Although the term has been used in other ways, its utility as an evolutionary concept in species interactions requires that the process involves partial co-ordination of non-mixing gene pools through reciprocal evolution (Thompson 1982). As Strong *et al.* (1984) put it, we are looking for coevolutionary vortices in an evolutionary stream. Showing mutual congruence of traits is not enough to suggest that coevolution has occurred (Janzen 1980). It follows that the term 'coadaptation' refers to reciprocal evolutionary adaptation in interacting species, while 'cospeciation' refers to reciprocal speciation as a result of interactions between species. The process of coevolution can be highly species specific or more diffuse (Gilbert 1975; Janzen 1980; Fox 1981; Futuyma and Slatkin 1983); in fact, through the combined processes of adaptation and speciation, coevolving interactions that are initially highly species-specific can become more diffuse over evolutionary time. As was first partially appreciated by nineteenth century biogeographers and is now much clearer, one long-term result may be partial convergence in the structure and organization of communities in similar physical environments (Orians and Paine 1983), since similar interactions often seem to arise in separate but similar environments (Thompson 1981).

Coevolution and systematics: Umbelliferae and insects

Since the title of this book is coevolution and systematics, we should consider the use of coevolution in the study of systematics and of systematics in the study of coevolution. As Stone (1984) noted, an appreciation of coevolution, especially of parasite–host interactions, is useful as an aid in systematics in sorting out the taxonomy of sibling species groups, population differentiation for resistance in their hosts, and the practical management of interactions. Conversely, the unique contribution of systematics in the study of coevolution is in establishing the patterns of relatedness among taxa that allow tests of hypotheses on the probability and direction of coevolution between species. Hypotheses on patterns in coevolution often take the following form:

Among closely-related species that differ in characteristic x, I hypothesize that coevolution (1) is more likely . . ., or (2) favours . . ., or (3) follows the following pattern.

Evaluation of hypotheses on patterns in coevolution that involve systematics requires a detailed understanding of (1) the systematics of both (or all) groups suggested to have undergone coevolution, (2) the biogeography of the interactions, and (3) the ecological outcomes of interactions between the species. An analysis and interpretation of coevolution can have different outcomes depending upon the degree of importance given to each of these components. The Ehrlich and Raven (1964) scenario of coevolution, which is a theory of the relationship of coevolution to systematics, is a formulation of a pattern in coevolution that requires an understanding of all these components.

Here the application of the Ehrlich–Raven scenario to the coevolution of the plant family Umbelliferae and the phytophagous insects that feed on umbellifers is considered to show how each of the above three factors is important for interpretation. Recently, Berenbaum (1983) suggested that these insects and plants have coevolved via changes in the coumarin chemistry of the plants in a way that fits the Ehrlich and Raven (1964) scenario. Her analysis suggested that coumarins have played a role in the evolution of these interactions. It is argued here, however, that the analysis does not provide a convincing case that coumarins have been a major force in generating patterns of adaptive radiation via the Ehrlich-Raven scenario if we consider the relationships among umbellifer taxa, the biogeography of the interactions, and the patterns of host use by the insects that feed on these plants.

The steps in the Ehrlich–Raven theory

(1) Plants produce novel secondary compounds through mutation and recombination. Berenbaum (1983) showed that coumarins have given rise to hydroxycoumarins, which have given rise to linear furanocoumarins (LFCs), which have in turn given rise to angular furanocoumarins (AFCs) (Fig. 7.1). This increasing complexity in coumarin chemistry is the basis for her subsequent arguments.

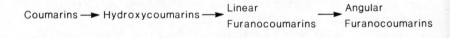

Coumarins	Hydroxycoumarins	Linear Furanocoumarins	Angular Furanocoumarins
33 families	31 families	8 families	2 families

Fig. 7.1. The distribution of different classes of coumarins among plant families; from Berenbaum's (1983) calculations. Plants produce novel secondary compounds through mutation and recombination.

(2) New secondary compounds alter the suitability of plants as food for insects. Berenbaum argued that LFCs are more toxic to polyphagous insects than hydroxycoumarins, and AFCs are more toxic to oligophagous insects (that feed on umbellifers) than linear furanocoumarins. The implication is that evolution of angular furanocoumarins in some umbellifer genera made those genera less suitable to oligophagous umbellifer-feeding insects, and she cited the low fecundity in one population of black swallowtails fed angelicin (an AFC) as evidence. The data from this one population, however, are an insufficient test. The *machaon* group of swallowtails, to which the black swallowtail (*Papilio polyxenes*) belongs, has populations throughout North America and Europe that feed regularly upon umbellifer genera that have AFCs: for example, *Papilio zelicaon* on *Angelica hendersonii*, *P. polyxenes* on *Pastinaca sativa*, and *P. machaon* on *A. archangelica*. Other *Papilio* populations within these species use hosts that lack furanocoumarins or have only linear furanocoumarins: *P. polyxenes* on *Daucus carota*, and

P. zelicaon on *Lomatium grayi* and *Cymopterus terebinthinus*. Such population differentiation may be similar to the adaptation of different populations of *Euphydryas editha* to different hosts (Singer 1971; Ehrlich and Murphy 1981; Rausher 1982). In fact, at least 5 of the 10 species of the *P. machaon* group in North America have some populations that feed upon AFC-containing umbellifers. Also, Wiklund's (1975) analysis of larval mortality on 57 umbellifers for *P. machaon* in Sweden, where it usually feeds on *A. archangelica*, shows no evidence that mortality is disproportionately higher on plants with AFCs. Hence, the conservative initial conclusion is that there is neither preference nor avoidance of AFC-containing plants *per se*. [Berenbaum (1983) noted later in her paper that the toxicity of AFCs to insects not affected by LFCs was currently speculative.]

(3) Plants with these new compounds undergo evolutionary radiation into a new adaptive zone. Berenbaum used the number of species per genus as an index of evolutionary radiation into a new adaptive zone. She suggested that the mean number of species per genus in umbellifers increases with increasing complexity of coumarins. She then argued that this greater number of species per genus is evidence that the development of angular furanocoumains resulted in adaptive radiation of those genera (Table 7.1).

Table 7.1. Percentage of apioid umbellifer genera listed in Berenbaum (1983) that are totally or partly European, for each coumarin category species/genus from Berenbaum (1983)

| Coumarins | Number | | % (N) in Flora Europaea |
	Genera	Species/genus $\pm$ S.D.	
None or dihydroxycoumarins	27	12.1 $\pm$ 3.6	48.1 (13)
Linear furanocoumarins	24	17.4 $\pm$ 5.8	66.7 (16)
Linear and angular furanocoumarins	11	67.3 $\pm$ 16.8	100.0 (11)

This is a difficult claim to evaluate because the analysis is confounded by geography and taxonomic traditions. Berenbaum used Drude's (1898) classification of genera in her analysis. Among those genera given in Berenbaum's Table 3, a disproportionate number of the large genera with AFCs are European or Eurasian (Table 7.1). The large number of species per genus in these European and Eurasian genera could have resulted from the presence of AFCs. However, such grouping of species could also have resulted from differences in geographical opportunities for speciation in the Old World as compared to the New World, differences in the absolute amount of land mass covered by the genera in the Old *v* New World, or differences in taxonomic traditions.

In fact, many of these large Old World genera are catch-all genera that need revision. As Heywood (1971) noted, generic classification in the Old World Umbelliferae is still in flux and 'several of the large genera may be unnatural'. Among the potentially unnatural groups, he mentioned specifically '*Seseli*, *Ligusticum*, *Pimpinella*, *Angelica*, *Daucus*, etc'. The first four of these genera are in the list of eleven that Berenbaum uses to argue that genera with AFCs are particularly large (Table 7.2). Similar doubt has been cast upon the large genus *Peucedanum* (*sensu lato*), which is also in the list of eleven cited genera with AFCs. Burtt and Davis (1949) noted that, 'In this group generic and subgeneric limits are greatly in need of redefinition'; Theobald (1971) complained that this genus is poorly understood and has never undergone a complete revision on a world-wide basis; and Constance and Chuang (1982) stated that within this genus 'problems of definition abound'. Some of the historically large European and Eurasian genera include groups from other geographical regions that undoubtedly belong in separate genera. For example, a number of New Zealand umbellifers were placed previously into the large genera *Angelica*, *Peucedanum*, and *Ligusticum*, which are all among the large genera with AFCs on Berenbaum's list. During the past 25 years, however, revisions of the New Zealand flora have moved these species into separate genera (Dawson 1961, 1971; Moar 1966; Dawson and Webb 1982). Hence, all but two of the large (>60 spp.) AFC-containing genera cited by Berenbaum are groups that systematists recognize as needing revision.

There is also still flux in the generic limits of New World genera, and

Table 7.2. Apioid umbellifer genera with linear and angular furanocoumarins listed in Berenbaum (1983)

Genus	Number of species*	Genus may be unnatural or needs redefinition
Bupleurum	150	
Pimpinella	150	Heywood (1971)
Peucedanum	120	Burtt and Davis (1949); Theobald (1971); Constance and Chuang (1982)
Seseli	80	Heywood (1971)
Angelica	80	Heywood (1971)
Heracleum	70	
Ligusticum	60	Heywood (1971)
Pastinaca	15	
Ammi	10	
Selinum	3	
Archangelica	2	Tutin *et al.* (1968) incl. in *Angelica*

*Numbers as given in Berenbaum (1983).

Mathias (1971) argued that a systematic revision is needed of generic groups on a worldwide basis. Some of the small North American genera that Berenbaum cited as lacking AFCs, have been combined with larger genera that lack AFCs since the publication of Drude's (1898) generic classification. For example, the small genus *Leptotaenia* has long been combined with the large genus *Lomatium* (Mathias and Constance 1942), and the small genus *Pteryxia* has been combined with *Cymopterus* by some authors (Hitchcock *et al.* 1968). Consequently, the major problems in the generic classification of umbellifers and the potential geographical differences in how species have been placed into genera preclude any assertion now that genera with AFCs are disproportionately large within the Umbelliferae.

(4) Insects evolve novel mechanisms of resistance to deal with these new compounds. Most of the insects that feed upon umbellifers are restricted to one or a few plant species, but it is not yet clear that coumarins are the major determinants of specificity. Berenbaum demonstrated that some insects can feed especially upon plants with AFCs. She also argued that leaf-tying in oecophorid moths preadapted these insects to radiate onto umbellifers with furanocoumarins because this behaviour allowed them to avoid the photoxic properties of some of these compounds. However, I am unaware of any studies showing that leaf-tying has any direct effect on the ability of insects to feed on plants containing potentially toxic furanocoumarins. Oecophorid moths are probably the most speciose group of insects that feed on umbellifers and tie leaves or umbels. Umbel-tying and leaf-tying are common in these genera, including species that feed on umbellifers that lack either LFCs or AFCs (Thompson 1983*a,b,c*). This common behaviour suggests that furanocoumarins are not, at least currently, the major selective force that maintains tying.

(5) The insects able to overcome these new chemical defences enter a new adaptive zone and radiate in species.

If Berenbaum's analysis of coumarin-induced plant speciation within the Umbelliferae is correct, then the evolution of LFCs had little effect on plant speciation, whereas the evolution of AFCs caused an explosion in species richness in those genera that contained them. (Her analysis showed a significant increase in the number of species per genus only with the evolution of AFCs.) Consequently, the subsequent radiation in insect species should be expected to be onto umbellifer genera containing AFCs if coevolution followed the Ehrlich–Raven scenario with coumarin evolution as the driving force. I do not believe that the evidence supports this claim.

Berenbaum used the Oecophoridae in North America and the Papilionidae world-wide as evidence for furanocoumarin-induced adaptive radiation, citing the number of species per genus or section of a genus. The analysis, however, is by genus and creates an unbalanced picture of the distribution of these insects on plant species: a whole insect genus is counted even if only a few species feed upon AFC-containing plants. Here the evidence for only the Oecophoridae is reconsidered, because the problem of population differentiation in the Papilionidae was considered above.

Berenbaum noted that the genera *Depressaria* and *Agonopterix* are the largest genera within the Depressariini (Oecophoridae), and it is primarily these two genera that feed upon plants containing furanocoumarins. Her implication is that most of the species in these genera feed upon plants containing AFCs, as is required by step 5 in the Ehrlich–Raven scenario.

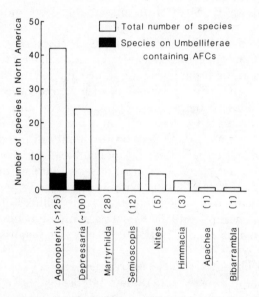

Fig. 7.2. The number of species per genus in North American Depressariini (Oecophoridae) and the number of species in each genus that feed on Umbelliferae containing AFCs. The worldwide number of species in each genus is given after the name of each genus. Data from Clarke (1952), Hodges (1974), Thompson (1983*a*, unpubl. data).

However, the evidence for the North American members of these genera is strongly to the contrary. In fact, among the 94 species of Depressariini in North America, only 8 (8.5 per cent) are known to feed on umbelliferous hosts that have AFCs (Fig. 7.2). Most North American *Depressaria* have radiated in species onto the genus *Lomatium* or *Cymopterus* (Clarke 1952; Hodges 1974; Thompson 1983*a,b,c*), both of which apparently lack LFCs and AFCs. Similarly, most species of

Agonopterix in North America feed either on trees or on umbellifers that lack LFCs or AFCs (Hodges 1974). The North American Depressariini comprise about one-third of the species world-wide. If the non-North American species feed disproportionately on AFC-containing genera, which is not yet known, then there will be at least two biogeographical and phylogenetic patterns to be explained in the evolution of these interactions.

I am not arguing against the importance of chemistry in the evolution of these interactions; plant chemistry has undoubtedly influenced the evolution of these interactions. Rather, the point is to argue that to consider a general hypothesis such as the Ehrlich–Raven scenario will require a detailed understanding of the systematics, biogeography, and natural history of these groups from a variety of perspectives in order to sort out the evolution of these interactions. The analysis of coumarin chemistry alone does not provide a convincing case for coevolution via the Ehrlich–Raven scenario. Recently, Smiley (1985) made similar arguments, questioning the coevolution of *Passiflora* vines and *Heliconius* butterflies via *Passiflora* chemistry.

Coevolution and speciation

1. Under what ecological conditions is cospeciation likely to occur?

Interactions between species may be among the most important causes of divergence of populations and eventual speciation (Bock 1972, 1979), and the Ehrlich-Raven scenario is only one of several kinds of ways coevolution affects speciation. Another way is cospeciation. Some kinds of interaction that have involved great radiations in species suggest long-term, diffuse coevolution, such as the evolution of grasses and grazing mammals (Stebbins 1981; Mack and Thompson 1982; McNaughton 1984). However, it seems unlikely in the evolution of mammal-grass interactions that the interaction itself was the cause of the repeated evolution of reproductive isolation in these taxa, and that is the crux of the problem: under what conditions does an interaction cause reproductive isolation of populations in interacting species?

Although some studies have shown parallel cladogenesis of interacting species (e.g. Mitter and Brooks 1983), there are currently few viable examples of cospeciation, that is reciprocal speciation induced by an interaction between two or more species. The two most likely cases are agaonid wasps and figs, and yucca moths and yuccas. However, these interactions are exceptional in that (1) they involve an interaction in which one species (the pollinator) has direct control over reproduction in the other species (the plant) and (2) these are rare kinds of pollinator–plant interactions in which the interaction appears to have evolved from a parasite–host interaction. Coevolution between pollinators and plants is probably generally more diffuse (Wheelwright and Orians 1982; Schemske 1983), especially now that there has been such a radiation of pollinator taxa such as bees, butterflies, and hummingbirds. As these pollinator taxa have evolved and radiated to exploit flowers, the world has become partly divided into habitats in which pollination by some taxa is more common than others.

Recently, O. Pellmyr (unpubl. data) found that *Cimicifuga simplex*, a Japanese herb, has three distinct floral morphs that are separated by elevation or habitat, and correspond to differences in visitation by pollinator taxa. At high elevations the plants are pollinated by short-tongued bumblebees. At low elevations the plants occur in two habitats. In sunny habitats, the plant is an inferior competitor with *Impatiens textori* for visits by bumblebees; *C. simplex* plants in these habitats produce a fragrance that is absent in the high elevation populations and is attractive to fritillary butterflies, the major pollinators of *C. simplex* individuals at these sunny sites. A third morph grows in shaded, moist, lowland habitats. These small, forest plants lack the fragrances found in the morph in the sunny habitats. These plants flower later in the season after *I. textori* has finished flowering and are pollinated by bumblebees. Incipient speciation in *C. simplex*, therefore, is occurring on the template of past radiation of pollinator taxa.

Cospeciation in parasites and hosts has also been difficult to decipher. In one of the most promising examples, Nault and DeLong (1980) suggested that *Dalbulus* leafhoppers may have cospeciated with maize and its ancestors via host race formation in the leafhoppers, and reproductive isolation in the plants mediated by spiroplasmas, mycoplasmas, and viruses. The leafhoppers transmit these diseases, which differ elevationally in their occurrence and vary in their virulence on different plant genotypes. Since then, Nault and colleagues have continued to sort out the complexity in the evolution of these interactions. Extensive collecting in North America and Mexico has shown that *Dalbulus* leafhoppers are narrowly oligophagous rather than monophagous (Nault *et al.* 1984); experiments on developmental rates have shown that the life histories of *Dalbulus* species are adapted to different life histories in *Zea* and *Tripsacum* (Nault and Madden 1985); and experiments on leafhopper survival have shown that pathogenicity of the spiroplasmas and mycoplasmas to the leafhoppers varies among leafhopper species (Madden and Nault 1983; Nault *et al.* 1984). These results show close adaptation of the insects to related taxa in *Tripsacum*-maize evolution and suggest evolution of the leafhoppers with the microorganisms they transmit; but the oligophagy of the leafhoppers also suggests that sorting out the reciprocal evolutionary effects will be complicated.

2. *How often is coevolution a process of adaptation in one species and speciation in the other species (mixed-process coevolution)?*

Coevolution need not, of course, be either coadaptation or cospeciation exclusively. Mixed-process coevolution may be defined as reciprocal evolution of interacting species in which adaptation of a population of one species to a population of a second species causes the population of the second species to become reproductively isolated from other populations (Fig. 7.3). Although this process clearly occurs as diffuse coevolution over long periods of evolutionary time, as specific coevolution it is probably limited mostly to interactions between species-specific pollinators and plants, parasite and hosts, and mutualistic symbionts and hosts.

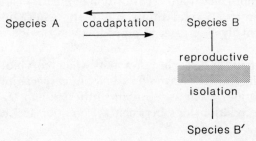

Fig. 7.3. Mixed-process coevolution, in which adaptation of one species to a population of a second species causes reproductive isolation in that population of the second species.

The kind of interaction that seems most likely to generate mixed-process coevolution between two species is one in which two conditions are met. Firstly, the parasite or mutualist becomes adapted to a particular population of a host and is not readily able to attack or use individuals in other populations of the host. Secondly, the parasite or mutualist has direct control over reproduction in the host and/or affects the viability of the host's offspring. Such direct control can occur under at least three kinds of situations: (1) a mutualist is directly responsible for transfer of the host's gametes as, for example, in pollinators of plants, (2) a maternally-inherited mutualistic symbiont is required by one host population, but not by other populations, or (3) a maternally-inherited commensal is highly virulent, or causes cellular or gametic disruption in other host populations.

Increasing evidence suggests that coevolution of insects and microorganisms has sometimes been through mixed-process coevolution. Reproductive isolation between populations and closely related insect species is sometimes caused by cytoplasmic incompatibility. Some of these cases of incompatibility seem to result from the presence of microorganisms that occur in the cytoplasm of one, but not the other host population and are transmitted maternally. Examples of such symbiont-induced reproductive isolation are now known from several orders of insects. They include *Culex pipiens* mosquitoes in which some populations harbour the rickettsia-like microorganism *Wolbachia pipientis* (Yen and Barr 1971, 1973); in *Drosophila paulistorum* in which mycoplasma-like symbionts are partly responsible for hybrid male sterility in crosses among semispecies (Ehrman and Kernaghan 1971; Ehrman and Powell 1982); the almond moth *Ephestia cautella* in which strains that harbour *Wolbachia* sp. cannot be crossed successfully with strains that do not harbour the symbiont (Kellen and Hoffman 1981); the alfalfa weevil *Hypera postica* in which rickettsia-induced incompatibility occurs among strains (Hsaio and Hsaio 1985); and *Tribolium confusum* in which microorganism-induced incompatibility occurs among some strains (Wade and Stevens 1984). Furthermore, germ-line transmission of micro-organisms has been demonstrated in a variety of animal and plant taxa (Mims 1981), and hence may be involved in cases of reproductive isolation of populations in other taxa.

There are, then, at least three ways in which coevolution is related to speciation: the Ehrlich–Raven formulation, cospeciation, and mixed-process coevolution. Each way presents different expectations on how to evaluate the relationships between systematics and coevolution of particular taxa.

Limits to the arms race and the evolution of mutualism

The variety of interactions that have coevolved form a continuum from antagonism to mutualism, and interactions may shift back and forth along this continuum over evolutionary time. Consequently, the commonly used arms race analogy of coevolution is an unduly confining view of coevolution for two reasons: it focuses too narrowly on one of several aspects of coevolution between antagonists, and it does not provide the perspective that is necessary to understand the evolutionary continuum between antagonism and mutualism. Here several overlapping hypotheses on patterns in coevolution that suggest the variety of coevolutionary outcomes that may occur along the antagonism–mutualism continuum are considered.

1. The arms race hypothesis

Long-term coevolution is through directional selection that regularly incorporates new mutations for defence and counter-defence in interacting species.

This hypothesis has appeared in several forms in evolutionary biology based upon a variety of ecological arguments (Dawkins and Krebs 1979). It is undoubtedly a common form of coevolution, but the conditions under which evolutionary arms races are fairly symmetric in selection pressures are only partially clear (Anderson and May 1979; Slatkin and Maynard Smith 1979). Moreover, as Levin and Lenski (1983) noted, lowered competitive ability in mutants and physiological constraints can decrease the likelihood of long-term arms race coevolution.

2. The polymorphism hypothesis

Arms race coevolution occurs most rapidly in new interactions; once a variety of alleles for defence and counter-defence have built up in the populations, most subsequent coevolution occurs through frequency-dependent selection, density-dependent selection, heterotic balance, and other forms of balancing selection, such as seasonally-dependent selection.

As new mutations are favoured in species early in the evolution of an interaction, polymorphisms in traits for defence and counterdefence can develop. Subsequently, much of coevolution probably becomes the continual fluctuation of allele frequencies and the addition of occasional new mutations. Consequently, the study of the selective value of a particular characteristic involved in coevolution is often not useful without a concomitant understanding of several related factors: the frequency of alternative characters within the population (frequency dependence), the outcomes of selection on the interaction at different population densities (density dependence) and at different times of year (seasonal dependence), and the selective value of the trait in different gene combinations (heterotic and pleiotropic effects). The evolution of the sickle-cell gene in humans shows how the outcomes for gene frequencies and the maintenance of polymorphism can vary with population structure (Templeton 1982). Haldane (1949) and Clarke (1976, 1979) suggested that many biochemical polymorphisms are probably maintained in populations through frequent-dependent selection in interactions between parasites and hosts.

Nonetheless, virtually no ecological experiments on natural populations have been designed to examine coevolution and polymorphism (i.e. testing both or all interacting species). As a result we are probably grossly underestimating the importance of coevolution between species as we concentrate on searching for novel traits in coadaptation and instances of cospeciation.

3. The decreasing effect hypothesis

Many interactions between species that differ greatly in generation time or size will decrease over time in likelihood of further coevolution, because the shorter lived and smaller species will be selected to attack the host (a) at late- or post-reproductive stages of the host's life history or (b) in ways that elicit less of a response from the host and have less of a selective effect on the host.

Late- or post-reproductive hosts may have less effective defences than pre-reproductive hosts, and these defences may favour especially parasites that selectively attack those older hosts or remain latent until the host is older (Thompson 1982). Similarly, a host could channel parasites onto body parts that are less detrimental to its fitness by defending body parts differentially (Thompson 1983a). Closely-related parasites sometimes differ in the parts of the hosts that they feed upon, so such variation may sometimes be present on which selection can act (Thompson 1983b,c). Through these processes the interaction could tend toward commensalism and decreased subsequent coevolution. The more similar the generation times of the host and parasite, and the more similar their sizes, however, the less likely it is that commensalism can occur through specialization on old hosts or relatively unimportant body parts.

4. The vector–virulence hypothesis

Evolution of parasite–host interactions towards commensalism or intermediate levels of severity is less likely when the parasite is transmitted by an intermediate vector species than when the parasite does not use a vector species.

Selection on parasites for reduced severity in interactions with hosts is probably effective only under a restrictive set of ecological and genetic conditions (Anderson and May 1980, 1982; Levin 1983; May 1983; May and Anderson 1983). Recently, Ewald (1983) argued that selection for reduced severity in infections will be less effective if the parasite is usually transmitted to new hosts by an intermediate vector. His preliminary correlational analysis showed that human diseases at or below the unicellular level of organization that are transmitted by arthropod vectors are significantly more lethal than diseases that are transmitted without vectors. It is unknown if selection on the host could differ in these different kinds of interactions or if variants of this intriguing idea could be expanded to some other kinds of parasite–host interaction.

5. The transition to mutualism hypothesis

Antagonistic interactions can become transformed over evolutionary time into mutualistic interactions in environments in which the benefit in fitness of the interaction outweighs the cost to the interacting species.

Mutualisms are common and are probably the very basis for eukaryotic life (Margulis 1981, 1984); in one laboratory colony parasitic bacteria of *Amoeba proteus* have even been observed to become symbionts that are required by the *Amoeba* (Jeon and Jeon 1976). Indeed, many, perhaps most, mutualisms originated as antagonistic interactions. There seem to be two major sets of ecological conditions that particularly favour coevolution in mutualism, one of which involves antagonism directly (Thompson 1982); many coevolved mutualisms appear to derive evolutionarily through a change in outcome in inevitable antagonistic interactions. Such a change may be most likely in organisms with intermediate survival ability (Roughgarden 1975, 1983) and in habitats with intermediate disturbance regimes (Thompson 1982). These are the kinds of ecological conditions in which the probability of interaction between species is high and slight inputs by a mutualist can have large effects on fitness. A slight shift in the mechanics of the interaction (Thompson 1985) could change an outcome from slightly antagonistic to slightly mutualistic. At the extremes of life histories and disturbance regimes, the evolution of mutualisms from antagonistic interactions is less likely because either the probability of encounter between the species will be too low or the input required by one species to have a mutualistic effect on the other species will be too large (Fig. 7.4).

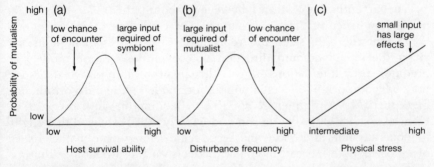

Fig. 7.4. Suggested probability of the evolution of mutualism with respect to three ecological axes: (a) survival ability of a host (Roughgarden 1975, 1983), (b) disturbance frequency, size, and intensity (Thompson 1982), and (c) physical (especially nutrient) stress (Thompson 1982).

The second set of ecological conditions that seems particularly to favour mutualism includes organisms with a high probability of encounter and very low premutualism growth rates in extreme environments that lack many antagonistic interactions (Thompson 1982) (Fig. 7.4). For example, mutualisms are commonly associated with nutrient-poor environments (Lewis 1973), and obligately mycorrhizal tropical plants reach dominance on nutrient poor habitats (Janos 1980) as do ant-fed plants (Janzen 1974; Huxley 1980; Thompson 1981). In these environments a small input by a mutualist could have large effects on fitness in the other species.

Population structure and coevolution

1. How does population structure affect coevolution?

Development of a firmer theoretical framework in the search for patterns in coevolution requires that we understand how population structure affects the direction of coevolution in interactions. Nonetheless, there are currently few studies that have addressed the problem except from the perspective of population size, density, or the frequency of polymorphic alleles within populations. Gilbert (1983) suggested that coevolution in Batesian mimicry is most likely only when the edible mimic is more abundant than the model (at least temporarily); coevolution in Mullerian mimicry is most likely when the two or more distasteful species have approximately the same population sizes, but fluctuate in relative abundances over time. In the coevolution of parasites and hosts, the studies of Anderson and May (e.g. Anderson and May 1982; May and Anderson 1983) have combined epidemiological approaches that involve population density with genetic studies to suggest conditions in which parasitism can evolve toward reduced virulence over evolutionary time.

We currently know little, however, about other aspects of how population structure affects coevolution. There are three major questions.

(1) How do age, size, and reproductive structures of populations affect the likelihood and direction of coevolution? Some models of predator–prey interactions suggest that incorporation of age structure may have important effects in population models (Maynard Smith and Slatkin 1973; Auslander *et al.* 1974; Beddington and Free 1976, Charlesworth 1980). Thompson and Moody (1985) showed that slight to moderate shifts in the size- and age-specific demographic characteristics of a host population can have large effects on the probability that an individual host will be attacked during its lifetime by a particular species of parasite. Furthermore, as populations change within patches over time in age, size, and reproductive structure (Grubb 1977; Thompson 1978, 1985; Price 1980; Pickett 1980), so can the outcomes of interactions. Consequently, there is a 'patch dynamics of interactions' as there is a patch dynamics of populations (Thompson 1982, 1985); the result can be repeated sequences of coevolution that result from predictable changes in population structure within patches over time.

(2) How does the social structure of populations influence the kinds of interspecific interactions that can occur and the outcome of coevolution? May (1983) hypothesized that directly transmitted microparasites should be more commonly associated with animals that live in herds or schools or breed in large colonies, since these kinds of parasites require high host densities in order to persist. I have argued that richness of social

behaviour and the evolution of mutualism are often positively associated (Thompson 1982).

(3) How does subdivision of populations into smaller demes influence coevolution and the persistence of an interaction? Black (1966) showed that measles does not persist in human populations on islands with less than 500 000 inhabitants: individuals acquire lifetime immunity after attack and, after an initial epidemic, there are too few hosts lacking immunity to maintain the parasite population. Recent models by Kiester et al. (1984) on pollinator–plant interactions suggest that rates of diversification during coevolution vary with the relative effective sizes of the populations. Furthermore, demic structure may generally influence the evolution of mutualism (Wilson 1983).

Interaction norms: a new term

Many of these suggested patterns in coevolution imply that the way in which selection acts on a coevolving interaction can vary with the physical and biotic environment in which the interaction takes place. The reaction norms of the genotypes (Schmalhausen 1949; Lewontin 1974; Gupta and Lewontin 1982) could generate different phenotypes in different environments and hence the outcomes of the interaction could differ greatly between populations. At the extreme, an interaction that is antagonistic in one population could be mutualistic in another population mostly because of reaction norms.

Here the term *interaction norm* is introduced, in parallel to reaction norm, as the variation in outcome of interactions across different environments as genotypes are held constant (Fig. 7.5).

Unfortunately, we have very few data on the mechanics and outcomes of interactions across a range of environments. McKey (1984; pers. comm.) found that the interaction between *Leonardoxa africana*, which is an understory tree of rainforest in west Africa, and ants varied geographically. Furthermore, characteristics of the plant important to the interaction, such as the size of the internodial swellings, and size and number of nectaries, varied geographically. Even fewer are studies on how interactions vary across environments if genotypes are held constant or are controlled. For example, Park's (1948, 1954) studies on *Tribolium confusum* and *T. castaneum* indicated that competitive outcomes could change as temperature and humidity changed. Bierbaum and Ayala (unpubl. manuscript) have shown that in the laboratory at 25°C *Drosophila willistoni* drives *D. pseudoobscura* to extinction, whereas at 18°C the converse occurs; coexistence can occur at intermediate temperatures. The outcome of interspecific competition in plants can vary with soil moisture (Pickett and Bazzaz 1978), nutrient ratios and levels (Tilman

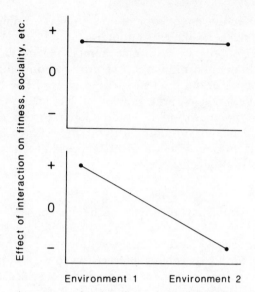

Fig. 7.5. Interaction norms, showing the variation in outcome of interactions across different physical or biotic environments as genotypes are held constant.

1982), and CO_2 levels (Bazzaz and Carlson 1984), and responses to herbivory can vary with the amount of shading (Dirzo 1984). These kinds of data are crucial if we are to evaluate the likelihood and potential directions of coevolution in different environments, and among related species that vary in life history and other characteristics.

The merging of approaches

Most of the questions and hypotheses posed here do not commonly appear in contemporary textbooks on evolution, systematics, ecology, or population genetics. Although particular problems in coevolution have now been studied for a couple of decades, it is only within the past decade that the breadth of questions we need to ask about coevolutionary change has become clear. In a review six years ago of models of coevolution, Slatkin and Maynard Smith (1979) could note that researchers had devoted little attention 'to the unique features of coevolution that could not arise from purely ecological or genetic models'. Since then, however, the situation has changed dramatically.

Increasingly coevolutionary studies merge approaches from a variety of disciplines: population ecology and population genetics (e.g. Leon 1974; Levin and Udoric 1977; Roughgarden 1977, 1979) or quantitative genetics (Bulmer 1974; Slatkin 1980; Taper and Case 1985); population biology and physiological ecology (Bazzaz 1984); systematics and ecology (Colwell and Winkler 1984); modes of interaction and quantitative genetics (Kiester *et al.* 1984); modes of interaction and population ecology (Thompson and Moody 1985); probability theory and community ecology (Simberloff 1980; Strong 1980; Harvey *et al.* 1983; Connor and Simberloff 1984; Grant and Schluter 1984); and game theory and behaviour (Lawlor and Maynard Smith 1976; Axelrod and Hamilton 1981; Maynard Smith 1982). This merging of approaches is necessary to consider the problems in coevolution that integrate studies of population biology, systematics, and geographical variation.

In summary, the questions that fuse these areas of study on coevolution are these: (1) how and when does coevolution affect speciation through the Ehrlich–Raven scenario, cospeciation, or mixed-process coevolution; (2) when is the arms race analogy applicable and when does coevolution result in changing modes of interaction; (3) how does population structure (age, size, social, demic) influence coevolution; and (4) how do the outcomes of interactions vary across environments (interaction norms)?

Acknowledgements

I thank D. J. Futuyma, R. N. Mack, J. M. Scriber, and D. R. Strong, Jr. for discussions or helpful comments on the manuscript and T. Bierbaum and O. Pellmyr for allowing me to cite unpublished manuscripts. I particularly thank May Berenbaum for good-naturedly discussing with me the similarities and dissimilarities in our perceptions of the current data on coevolution of umbellifers and insects. I am grateful to D. Davis, P. Marsh, and L. R. Nault for conversations on the evolution of specificity in insects. This work was supported in part by NSF grant BSR 8219884 and USDA (Biological Stress Competitive Grant) 84-CRCR-1-1395.

References

Anderson, R. M., and May, R. M. (1979). Population biology of infectious diseases: Part I. *Nature, Lond.* 280, 361–7.
—— and —— (1980). The population dynamics of microparasites and their invertebrate hosts. *Phil. Trans. Roy. Soc. Lond., B* 291, 451–524.
—— and —— (1982). Coevolution of hosts and parasites. *Parasitology* 85, 411–26.
Auslander, D., Oster, G. F., and Huffaker, C. (1974). Dynamics of interacting populations. *J. Franklin Inst.* 297, 345–75.
Axelrod, R., and Hamilton, W. D. (1981). The evolution of cooperation. *Science*, N.Y. 211, 1390–6.

Bazzaz, F. A. (1984). Demographic consequences of plant physiological traits: some case studies. In *Perspectives on Plant Population Ecology* (eds R. Dirzo and J. Sarukhan), pp. 324-46. Sinauer Associates, Sunderland, Massachusetts.
—— and Carlson, R. W. (1984). The response of plants to elevated CO_2. I. Competition among an assemblage of annuals at two levels of soil moisture. *Oecologia* 62, 196-8.
Beddington, J. R. and Free, C. A. (1976). Age structure effects in predator-prey interactions. *Theor. Pop. Biol.* 9, 15-24.
Berenbaum, M. (1983). Coumarins and caterpillars: a case for coevolution. *Evolution* 37, 163-79.
Black, F. L. (1966). Measles endemicity in insular populations: critical community size and its evolutionary implications. *J. Theor. Biol.* 11, 207-11.
Bock, W. J. (1972). Species interactions and macroevolution. *Evol. Biol.* 5, 1-24.
—— (1979). The synthetic explanation of macroevolutionary change — a reductionistic approach. *Bull. Carnegie Museum Nat. Hist.* 13, 20-69.
Bulmer, M. G. (1974). Density-dependent selection and character displacement. *Am. Naturalist* 108, 45-58.
Burtt, B. L. and Davis, P. H. (1949). *Glaucosciadium*: a new Mediterranean genus of Umbelliferae. *Kew Bull.* 2, 225-30.
Charlesworth, B. (1980). *Evolution in Age-structured Populations*. Cambridge University Press, Cambridge.
Clarke, B. C. (1976). The ecological genetics of host-parasite relationships. In *Genetic Aspects of Host-parasite Relationships* (eds A. E. R. Taylor and R. Muller), pp. 87-103. Blackwell, Oxford.
—— (1979). The evolution of genetic diversity. *Proc. R. Soc. Lond. B* 205, 453-74.
Clarke, J. F. G. (1952). Host relationships of moths of the genera *Depressaria* and *Agonopterix*, with descriptions of new species. *Smithsonian Miscellaneous Collection* 117(7), 1-20.
Colwell, R. K. and Winkler, D. W. (1984). A null model for null models in biogeography. In *Ecological Communities: Conceptual Issues and the Evidence* (eds D. R. Strong, Jr., D. Simberloff, L. G. Abele, and A. B. Thistle), pp. 344-59. Princeton University Press, Princeton, New Jersey.
Connor, E. F. and Simberloff, D. (1984). Neutral models of species' co-occurrence patterns. In *Ecological Communities: Conceptual Issues and the Evidence*. (eds D. R. Strong, Jr., D. Simberloff, L. G. Abele, and A. B. Thistle), pp. 316-31. Princeton University Press, Princeton, New Jersey.
Constance, L. and Chuang, T.-I. (1982). Chromosome numbers of Umbelliferae (Apiaceae) from Africa south of the Sahara. *Bot. J. Linn. Soc.* 85, 195-208.
Dawkins, R. and Krebs, J. (1979). Arms races between and within species. *Proc. R. Soc. Lond., B* 205, 489-511.
Dawson, J. W. (1961). A revision of the genus *Anistome*. *Univ. Calif. Publ. Bot.* 33, 1-98.
—— (1971). Relationships of the New Zealand umbelliferae. In *The Biology and Chemistry of the Umbelliferae* (ed. V. H. Heywood), pp. 43-61. Academic Press, New York.
—— and Webb, C. J. (1982). Generic problems in Australasian Apioideae (Umbelliferae). *Monogr. Syst. Bot.* 6, 21-32.

Dirzo, R. (1984). Herbivory: a phytocentric overview. In *Perspectives on Plant Population Ecology* (eds R. Dirzo and J. Sarukhan), pp. 141-65. Sinauer Associates, Sunderland, Massachusetts.

Drude, C. G. O. (1898). Umbelliferae. In *Die natürlichen Pflanzenfamilien* (eds A. Engler and K. Prantl), vol. 3, pp. 63-250. Duncker and Humboldt, Berlin.

Ehrlich, P. R. and Murphy, D. D. (1981). The population biology of checkerspot butterflies (*Euphydryas*). *Biol. Zbl.* 100, 613-29.

—— and Raven, P. H. (1964). Butterflies and plants: a study in coevolution. *Evolution* 18, 586-608.

Ehrman, L. and Kernaghan, R. P. (1971). Microorganismal basis of infections hybrid male sterility in *D. paulistorum*. *J. Hered.* 62, 67-71.

—— and Powell, J. R. (1982). The *Drosophila willistoni* species group. In *The Genetics and Biology of Drosophila*, Vol. 3b (eds M. Ashburner, H. L. Carson, and J. N. Thompson, Jr.), pp. 193-225. Academic Press, New York.

Ewald, P. W. (1983). Host-parasite relations, vectors, and the evolution of disease severity. *Ann. Rev. Ecol. Syst.* 14, 465-85.

Fox, L. R. (1981). Defense and dynamics in plant-herbivore systems. *Am. Zool.* 21, 853-64.

Futuyma, D. J. and Slatkin, M. (1983). Introduction. In *Coevolution* (eds D. J. Futuyma and M. Slatkin), pp. 1-13. Sinauer Associates, Sunderland, Massachusetts.

Gilbert, L. E. (1975). Ecological consequences of a coevolved mutualism between butterflies and plants. In *Coevolution of Animals and Plants*. (eds L. E. Gilbert and P. H. Raven), pp. 210-40. University of Texas Press, Austin, Texas.

—— (1983). Coevolution and mimicry. *Coevolution*. (eds D. J. Futuyma and P. H. Raven), pp. 263-81. Sinauer Associates, Sunderland, Massachusetts.

Grant, P., and Schluter, D. (1984). Interspecific competition inferred from patterns of guild structure. In *Ecological Communities: Conceptual Issues and the Evidence* (ed. D. R. Strong, Jr., D. Simberloff, L. G. Abele, and A. B. Thistle), pp. 201-33. Princeton University Press, Princeton, New Jersey.

Grubb, P. J. (1977). The maintenance of species richness in plant communities: the importance of the regeneration niche. *Biol. Rev.* 52, 107-45.

Gupta, A. P. and Lewontin, R. C. (1982). A study of reaction norms in natural populations of *Drosophila pseudoobscura*. *Evolution* 36, 934-948.

Haldane, J. B. S. (1949). Disease and evolution. *Ricerca Scient.* Suppl. 19, 68-76.

Harvey, P. H., Colwell, R. K., Silvertown, J. W. and Marg, R. M. (1983). Null models in Ecology. Ann. Rev. Ecol. Syst. 14. 189-211.

Heywood, V. H. (1971). Systematic survey of old World Umbelliferae. In *The Biology and Chemistry of the Umbelliferae* (ed. V. H. Heywood), pp. 31-49. Academic Press, New York.

Hitchcock, C. L., Cronquist, A., Ownbey, M. and Thompson, J. W. (1968). *Vascular Plants of the Pacific Northwest*. University of Washington Press, Seattle, Washington.

Hodges, R. W. (1974). Gelechoidea: Oecophoridae. In *The Moths of America North of Mexico* (eds R. B. Dominick, D. C. Ferguson, J. F. Franclemont, R. W. Hodges, and E. G. Munroe) Fasc. 6.2. Classey, London.

Hsaio, T. H. and Hsaio, C. (1985). Hybridization and cytoplasmic incompatibility among alfalfa weevil strains. *Entomol. Exp. Appl.* 37, 155-9.

Huxley, C. (1980). Symbiosis between ants and epiphytes. *Biol. Rev.* 55, 321-40.
Janos, D. P. (1980). Mycorrhizae influence tropical succession. *Biotropica* 12, (supplement), 56-64.
Janzen, D. H. (1974). Epiphytic myrmecophytes in Sarawak: mutualism through the feeding of plants by ants. *Biotropica* 6, 237-59.
—— (1980). When is it coevolution? *Evolution* 34, 611-12.
Jeon, K. W. and Jeon, M. S. (1976). Endosymbiosis in amoebae: recently established endosymbionts have become required cytoplasmic components. *J. Cell Physiol.* 89, 337-44.
Kellen, W. R. and Hoffmann, D. F. (1981). *Wolbachia* sp. (Rickettsiales: Rickettsiaceae) a symbiont of the almond moth, *Ephestia cautella*: ultrastructure and influence on host fertility. *J. Invert. Pathol.* 37, 273-83.
Kiester, A. R., Lande, R., and Schemske, D. W. (1984). Models of coevolution and speciation in plants and their pollinators. *Am. Naturalist* 124, 220-43.
Lawlor, L. R. and Maynard Smith, J. (1976). The coevolution and stability of competing species. *Am. Naturalist* 110, 9-99.
Leon, J. A. (1974). Selection in contexts of interspecific competition. *Am. Naturalist* 108, 39-57.
Levin, B. R. and Lenski, R. (1983). Coevolution in bacteria and their viruses and plasmids. In *Coevolution* (eds D. J. Futuyma and M. Slatkin), pp. 99-127. Sinauer Associates, Sunderland, Massachusetts.
Levin S. A (1983). Some approaches to the modelling of coevolutionary interactions. In *Coevolution* (ed. M. H. Nitecki), pp. 21-65. University of Chicago Press.
—— and Udovic, J. D. (1977). A mathematical model of coevolving populations. *Am. Naturalist* 111, 657-75.
Lewis, D. H. (1973). The relevance of symbiosis to taxonomy and ecology, with particular reference to mutualistic symbioses and the exploitation of marginal habitats. In *Taxonomy and Ecology* (ed. V. H. Heywood), pp. 151-72. Academic Press, New York.
Lewontin, R. C. (1974). The analysis of variance and the analysis of causes. *Am. J. Hum. Gen.* 26, 400-11.
Mack, R. N. and Thompson, J. N. (1982). Evolution in steppe with few large hooved mammals. *Am. Naturalist* 119, 57-73.
McKey, D. (1984). Interaction of the ant-plant *Leonardoxa africana* (Caesalpiniaceae) with its obligate inhabitants in a rainforest in Cameroon. *Biotropica* 16, 81-99.
McNaughton, S. J. (1984). Grazing lawns: animals in herds plant form, and coevolution. *Am. Naturalist* 124, 863-86.
Madden, L. V. and Nault, L. R. (1983). Differential pathogenicity of corn stunting mollicutes to leafhopper vectors in *Dalbulus* and *Baldulus* species. *Phytopathology* 73, 1608-14.
Margulis, L. (1981). *Symbiosis in Cell Evolution*. Freeman, San Francisco.
—— (1984). *Early Life*. Jones and Bartlett, Boston.
Mathias, M. E. (1971). Systematic survey of New World Umbelliferae. In *The Biology and Chemistry of the Umbelliferae* (ed. V. H. Heywood), pp. 13-29. Academic Press, New York.
—— and Constance, L. (1942). New combinations and new names in the Umbelliferae-II. *Bull. Torrey Bot. Club* 69, 244-8.

May, R. M. (1983). Parasitic infections as regulators of animal populations. *Am. Sci.* 71, 36-44.
—— and Anderson, R. M. (1983). Epidemiology and genetics in the coevolution of parasites and hosts. *Proc. Roy. Soc. Lond. B* 219, 281-313.
Maynard Smith, J. (1982). *Evolution and the Theory of Games*. Cambridge University Press, Cambridge.
—— and Slatkin, M. (1973). The stability of predator-prey systems. *Ecology* 54, 384-91.
Mims, C. A. (1981). Vertical transmission of viruses. *Microbiol. Rev.* 45, 267-86.
Mitter, C. and Brooks, D. R. (1983). Phylogenetic aspects of coevolution. In *Coevolution* (eds D. J. Futuyma and M. Slatkin), pp. 65-98. Sinauer Associates, Sunderland, Massachusetts.
Moar, N. T. (1966). Studies in pollen morphology. 3. The genus *Gingidium* J. R. and G. Forst in New Zealand. *N. Z. J. Bot.* 4, 322-32.
Nault, L. R. and DeLong, D. M. (1980). Evidence for coevolution of leafhoppers in the genus *Dalbulus* (Cicadellidae: Homoptera) with maize and its ancestors. *Annls Ent. Soc. Am.* 73, 349-53.
—— and Madden, L. V. (1985). Ecological strategies of *Dalbulus* leafhoppers. *Ecol. Entomol.* 10, 57-63.
—— —— Styer, W. E., Triplehorn, B. W., Shambaugh, G. F. and Heady, S. E. (1984). Pathogenicity of corn stunt spiroplasma and maie bushy stunt mycoplasma to their vector, *Dalbulus longulus*. *Phytopathology* 74, 977-9.
Orians, G. H. and Paine, R. T. (1983). Convergent evolution at the community level. In *Coevolution* (eds D. J. Futuyma and M. Slatkin), pp. 431-58. Sinauer Associates, Sunderland, Massachusetts.
Park, T. (1948). Experimental studies of interspecies competition I. Competition between populations of the flour beetles *Triolium confusum* and *Tribolium castaneum* Herbst. *Ecol. Monogr.* 18, 265-307.
—— (1954). Experimental studies of interspecies competition II. Temperature, humidity and competition in two species of *Tribolium*. *Physiol. Zool.* 27, 177-238.
Pickett, S. T. A. (1980). Non-equilibrium coexistence in plants. *Bull. Torrey Bol. Club* 107, 238-48.
—— and Bazzaz, F. A. (1978). Organization of an assemblage of early successful species on a soil moisture gradient. *Ecology* 59, 1248-1255.
Price, P. W. (1980). *Evolutionary Biology of Parasites*. Princeton University Press, Princeton, New Jersey.
Rausher, M. D. (1982). Population differentiation in *Euphydryas editha* butterflies: larval adaptation to different hosts. *Evolution* 36, 581-90.
Roughgarden, J. (1975). Evolution of marine symbiosis: a simple cost-benefit model. *Ecology* 56, 1201-8.
—— (1977). Coevolution in ecological systems: results from 'loop analysis' for purely density-dependent coevolution. In *Measuring Selection in Natural Populations*. (eds F. B. Christiansen and T. M. Fenchel), pp. 449-517. Lecture Notes in Biomathematics, vol. 19. Springer-Verlag, New York.
—— (1979). *Theory of Population Genetics and Evolutionary Ecology: An Introduction*. Macmillan, New York.

—— (1983). The theory of coevolution. In *Coevolution* (eds D. J. Futuyma and M. Slatkin), pp. 33–64. Sinauer Associates, Sunderland, Massachusetts.
Schemske, D. W. (1983). Limits to specialization and coevolution in plant-animal mutualisms. In *Coevolution* (ed. M. H. Nitecki), pp. 67–109. University of Chicago Press, Chicago.
Schmalhausen, I. I. (1949). *Factors of Evolution*. Blakeston, Philadelphia.
Simberloff, D. (1980). A succession of paradigms in ecology: essentialism to materialism to probabilism. *Synthese* 43, 3–39.
Singer, M. C. (1971). Evolution of food-plant preference in the butterfly *Euphydryas editha*. *Evolution* 25, 383–389.
Slatkin, M. (1980). Ecological character displacement. *Ecology* 61, 163–77.
—— and Maynard Smith, J. (1979). Models of coevolution. *Q. Rev. Biol.* 54, 233–63.
Smiley, J. T. (1985). Are chemical barriers necessary for evolution of butterfly–plant associations? *Oecologia* 65, 580–3.
Stebbins, G. L. (1981). Coevolution of grasses and herbivores. *Annls Missouri Bot. Gard.* 68, 75–86.
Stone, A. R. (1984). Changing approaches in nematode taxonomy. *Plant Disease* 68, 551–4.
Strong, D. R. (1980). Null models in ecology. *Synthese* 43, 271–85.
—— (1984). Exorcising the ghost of competition past: phytophagous insects. In *Ecological Communities: Conceptual Issues and the Evidence*. (eds D. R. Strong, Jr., D. Simberloff, L. G. Abele, and A. B. Thistle), pp. 28–41. Princeton University Press.
—— Lawton, J. H. and Southwood, R. (1984). *Insects on Plants: Community Patterns and Mechanisms*. Blackwell Scientific, Oxford.
Taper, M. L. and Case, T. J. (1985). Quantitative genetic models for the coevolution of character displacement. *Ecology* 66, 355–71.
Templeton, A. R. (1982). Adaptation and the integration of evolutionary forces. In *Perspectives on Evolution* (ed. R. Milkman), pp. 15–31. Sinauer Associates, Sunderland, Massachusetts.
Theobald, W. L. (1971). Comparative anatomical and developmental studies in the Umbelliferae. In *The Biology and Chemistry of the Umbelliferae* (ed. V. H. Heywood), pp. 179–97. Academic Press, New York.
Thompson, J. N. (1978). Within-patch structure and dynamics in *Pastinaca* and resource availability to a specialized herbivore. *Ecology* 59, 443–8.
—— (1981). Reversed animal–plant interactions: the evolution of insectivorous and ant-fed plants. *Biol. J. Linn. Soc.* 16, 147–55.
—— (1982). *Interaction and Coevolution*. Wiley-Interscience, New York.
—— (1983a). Selection pressures on phytophagous insects feeding on small host plants. *Oikos* 40, 438–44.
—— (1983b). Selection of plant parts by *Depressaria multifidae* (Lep., Oecophoridae) on its seasonally-restricted hostplant, *Lomatium grayi* (Umbelliferae). *Ecol. Entomol.* 8, 203–11.
—— (1983c). The use of ephemeral plant parts on small host plants: how *Depressaria leptotaeniae* (Lepidoptera: Oecophoridae) feeds on *Lomatium dissectum* (Umbelliferae). *J. Anim. Ecol.* 52, 281–91.

—— (1985). Within-patch dynamics of life histories, populations, and interactions: selection overtime in small spaces. In *The Ecology of Natural Disturbance* (ed. S. T. A. Pickett and P. S. White), pp. 253-64. Academic Press, New York.

—— and Moody, M. E. (1985). Assessing probability of interaction in size-structured populations: *Depressaria* attack on *Lomatium*. *Ecology* 66, 1597-607.

Tilman, D. (1982). *Resource Competition and Community Structure*. Princeton University Press, Princeton, New Jersey.

Tutin, T. G., Heywood, V. H., Burges, N. A., Moore, D. M., Valentine, D. H., Walters, S. M., and Webb, D. A. (eds) (1968). *Flora Europaea*. Vol. 2. *Rosaceae to Umbelliferae*. Cambridge University Press, Cambridge.

Wade, M. J. and Stevens, L. (1984). Microorganism mediated reproductive isolation in flour beetles (genus *Tribolium*). *Science*, N.Y. 227, 527-28.

Wheelwright, N. T. and Orians, G. H. (1982). Seed dispersal by animals contrasts with pollen dispersal, problems of terminology, and constraints on coevolution. *Am. Naturalist* 119, 402-13.

Wiklund, C. (1975). The evolutionary relationship between adult oviposition preferences and larval host plant range in *Papilio machaon* L. *Oecologia* 18, 185-97.

Wilson, D. S. (1983). The effect of population structure on the evolution of mutualism: a field test involving burjing beetles and their phoretic mites. *Am. Naturalist* 121, 851-70.

Yen, J. H. and Barr, A. R. (1971). New hypothesis of the cause of cytoplasmic incompatibility in *Culex pipiens* L. *Nature, Lond.* 232, 657-8.

—— and —— (1973). The etiological agent of cytoplasmic incompatibility in *Culex pipiens*. *J. Invert. Pathol.* 22, 242-50.

Systematics Association Publications

1. Bibliography of key works for the identification of the British fauna and flora *3rd edition* (1967)
   Edited by G. J. Kerrich, R. D. Meikle and N. Tebble
   Out of print
2. Function and taxonomic importance (1959)
   Edited by A. J. Cain
   Out of print
3. The species concept in palaeontology (1956)
   Edited by P. C. Sylvester-Bradley
   Out of print
4. Taxonomy and geography (1962)
   Edited by D. Nichols
   Out of print
5. Speciation in the sea (1963)
   Edited by J. P. Harding and N. Tebble
   Out of print
6. Phenetic and phylogenetic classification (1964)
   Edited by V. H. Heywood and J. McNeill
   Out of print
7. Aspects of Tethyan biogeography (1967)
   Edited by C. G. Adams and D. V. Ager
   Out of print
8. The soil ecosystem (1969)
   Edited by H. Sheals
   Out of print
9. Organisms and continents through time (1973)[†]
   Out of print
   Edited by N. F. Hughes
   Out of print

Published by the Association

Systematics Association Special Volumes

1. The new systematics (1940)
   Edited by Julian Huxley (Reprinted 1971)
   Out of print
2. Chemotaxonomy and serotaxonomy (1968)*
   Edited by J. G. Hawkes
3. Data processing in biology and geology (1971)*
   Edited by J. L. Cutbill
4. Scanning electron microscopy (1971)*
   Edited by V. H. Heywood
5. Taxonomy and ecology (1973)*
   Edited by V. H. Heywood
6. The changing flora and fauna of Britain (1974)*
   Edited by D. L. Hawksworth
7. Biological identification with computers (1975)*
   Edited by R. J. Pankhurst
8. Lichenology: progress and problems (1976)*
   Edited by D. H. Brown, D. L. Hawksworth and R. H. Bailey
9. Key works to the fauna and flora of the British Isles and north-western Europe (1978)*
   Edited by G. J. Kerrich, D. L. Hawksworth and R. W. Sims
   Out of print
10. Modern approaches to the taxonomy of red and brown algae (1978)*
    Edited by D. E. G. Irvine and J. H. Price
11. Biology and systematics of colonial organisms (1979)*
    Edited by G. Larwood and B. R. Rosen
12. The origin of major invertebrate groups (1979)*
    Edited by M. R. House
13. Advances in bryozoology (1979)*
    Edited by G. P. Larwood and M. B. Abbot
14. Bryophyte systematics (1979)*
    Edited by G. C. S. Clarke and J. G. Duckett
15. The terrestrial environment and the origin of land vertebrates (1980)*
    Edited by A. L. Panchen
16. Chemosystematics: principles and practice (1980)*
    Edited by F. A. Bisby, J. G. Vaughan and C. A. Wright
17. The shore environment: methods and ecosystems (2 Volumes) (1980)*
    Edited by J. H. Price, D. E. G. Irvine and W. F. Farnham
18. The Ammonoidea (1981)*
    Edited by M. R. House and J. R. Senior

19. Biosystematics of social insects (1981)*
    Edited by P. E. Howse and J.-L. Clément
20. Genome evolution (1982)*
    Edited by G. A. Dover and R. B. Flavell
21. Problems of phylogenetic reconstruction (1982)*
    Edited by K. A. Joysey and A. E. Friday
22. Concepts in nematode systematics (1983)*
    Edited by A. R. Stone, H. M. Platt and L. F. Khalil
23. Evolution, time and space: the emergence of the biosphere (1983)*
    Edited by R. W. Sims, J. H. Price and P. E. S. Whalley
24. Protein polymorphism: adaptive and taxonomic significance (1983)*
    Edited by G. S. Oxford and D. Rollinson
25. Current concepts in plant taxonomy (1984)*
    Edited by V. H. Heywood and D. M. Moore
26. Databases in systematics (1984)*
    Edited by R. Allkin and F. A. Bisby
27. Systematics of the green algae (1984)*
    Edited by D. E. G. Irvine and D. M. John
28. The origins and relationships of lower invertebrates (1985)[‡]
    Edited by S. Conway Morris, J. D. George, R. Gibson and H. M. Platt
29. Infraspecific classification of wild and cultivated plants (1986)[‡]
    Edited by B. T. Styles
30. Biomineralization in lower plants and animals (1986)[‡]
    Edited by B. S. C. Leadbeater and R. Riding
31. Systematic and taxonomic approaches in palaeobotany (1986)[‡]
    Edited by R. A. Spicer and B. A. Thomas
32. Coevolution and systematics (1986)[‡]
    Edited by A. R. Stone and D. L. Hawksworth

*Published by Academic Press for the Systematics Association
[†]Published by the Palaeontological Association in conjunction with the Systematics Association
[‡]Published by Oxford University Press for the Systematics Association